THE TOURMALINE

THE HISTORY OF MOUNT MICA OF MAINE, U. S. A.

Augustus Choate Hamlin, M.D.

Coachwhip Publications
Landisville, Pennsylvania

The Tourmaline / The History of Mount Mica of Maine, U. S. A.
Copyright © 2009 Coachwhip Publications

ISBN 1-930585-91-8
ISBN-13 978-1-930585-91-1

Coachwhipbooks.com

All Rights Reserved. No part of this publication may be reproduced, stored in a retrieval system or transmitted in any form or by any means—electronic, mechanical, photocopy, recording or any other—except for brief quotations in printed reviews, without the prior permission of the author or publisher.

The Tourmaline

Introduction	7
Chapter I	11
Chapter II	17
Chapter III	26
Chapter IV	52
Chapter V	70

The History of Mount Mica

Chapter I	77
Chapter II	91
Chapter III	101
Chapter IV	109
Chapter V	115
Chapter VI	118
Chapter VII	128
Plates	141

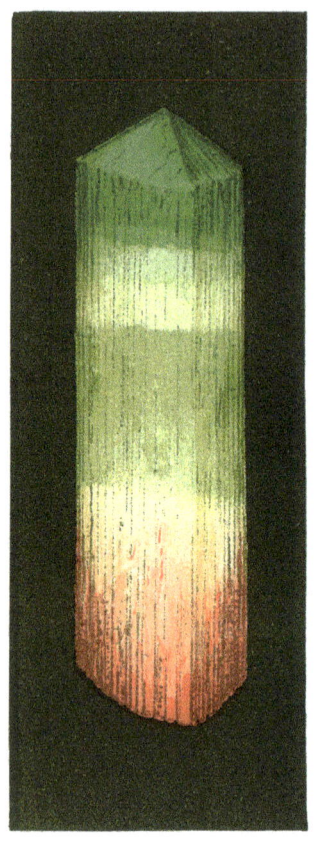

Crystal of Tourmaline
Exact Size.

Hamlin Collection.   Mt. Mica, ME

The Tourmaline

Its Relation as a Gem; Its Complex Nature;
Its Wonderful Physical Properties, etc., etc.;
With Special Reference to the Beautiful
and Matchless Crystals Found in the
State of Maine.

A. C. Hamlin, M.D.

"Membre de la Société Royale des Antiquaires du Nord,"
Member of the Academy of Sciences, Philadelphia, &c., &c.

(1873)

"It is a strange analogy, well worthy of fixing the attention of philosophers.... These jewels, which have the privilege of attracting our gaze, and of fixing our eyes upon them by an unaccountable species of magnetism, appear also to incite the secret affinities of lightning."—Abbé Fonvielle.

Introduction

The study of the gems and the precious stones furnishes many instructive and interesting themes to the student of science, the philosopher, or the historian. To either, these inanimate objects, which the learned Abbé Haüy elegantly called "the flowers of the mineral kingdom," afford a vast and almost limitless field of inquiry and research. The student, startled or entranced by the marvellous revelations of their physical properties, is led by his inquiries into the very depths of Nature's mysteries; and the historian, as he recalls the pages of the world's history, and seeks to fathom the strange fascinations which the gems have in all times exercised over the minds of men, listens with a well-defined feeling of credulity to the strange tales of Oriental fable. The visions of the Genii, the gigantic phantoms of the Afrites, are not so far distant and inexplicable, after all, when we come to consider the causes which give birth to the fabulous, or create in the bosom of the earth the transparent crystals of precious stone. The same telluric magnetism, or electricity, which elevates to the very heavens those lofty columns and weird-like forms of dust, or gives rise to the startling mirage of the horizon and the strange flashings and coruscations of light that appear in the darkness of night, also silently deposits in the earth, or even in the interior of the solid rocks, those wonderful crystals of symmetrical form and dazzling colors which are the very emblems of purity and perfection on earth. The philosopher, viewing the subject in an impartial light, will admit that this passion for the glittering and mysterious gems is not merely a love of finery, or an acquired taste dependent upon the freaks of fashion, but that it springs from a deeper source, and is inherent in human nature. It is exemplified in the rich grandee when lavishing millions for gem decorations, as well as in the child when gathering flowers on the meadow in spring to adorn its person. It is simply obedience to one of the laws that perfect our nature,—the love of the beautiful. We also find that the precious stones have, from time immemorial, exercised a powerful influence over all the human families, though widely scattered over the globe, not only in civilized and refined life, but among the wild Arabs of the desert, and the ruder savages of remote islands and secluded countries. How many beautiful examples of their effects come to us through the mists of antiquity, tinged, perhaps, with the rosy tints of the fabulous, but illustrating the force and sway of the beautiful in nature upon the disposition of man! How brilliant and potent must have been the gleams of that wonderful opal which induced the Roman senator Nonias to depart into exile rather than yield up his matchless gem to the greedy Marc Antony who coveted it! for exile then to a Roman was worse than death. Of what transcendent beauty must have been the iridized tints of that idolized pearl which the mediaeval Greeks lost to Alp Arslan, and which the Byzantine historians long lamented even more bitterly than all the Asiatic provinces won from them by the Turk! The traditions of thousands of years have given to many of the gems a glorious prestige, and have almost justified a belief in their marvellous properties. Most of these traditions and legends relating to the precious stones have come from Arabia and Persia,—those countries which are but stretches of rock and sand, yet whose every feature predisposes man to thought; where the visible is viewed only through the mirage of the imagination; and where the limpidity of the firmament, the serenity of the stars, and the transparency of the night, insensibly lead men to contemplation.

In these countries the superstitious mind is ever prone to seize upon the strange in nature, and transform the appearance or the action to the presence or power of some mighty and unseen force. Those who have travelled over the Arabian deserts, or across the almost limitless expanse of sands of Africa, can readily comprehend the origin of some of the creations that people Oriental fable and history. Those tall sand-pillars which mysteriously arise from the desert, and fairly pierce the vault of heaven with their awful forms,—moving majestically over the earth with a silent yet terrible power,—may readily suggest to human fancy the ideas of the gigantic Afrites; and, in like manner, many of the Arabian creations may be ascribed to an exalted and perverted view of natural phenomena. It is also not only interesting, but highly instructive, to trace out the origin of the mystic power of gems portrayed all through the charming tales of Eastern romance. We may then understand how this belief of magic and talismans, arising from the credulous simplicity of the shepherds of the plains and the deserts, afterwards passed to Greece and Rome, and gave birth to those exquisite forms of beauty and sentiment of the Greek and Roman art which have descended to us in the shape of engraved gems. There is yet a deeper and nobler interest in this subject, which commerce has debased in pandering to the frivolities of mankind; and it is impossible to study with untrammelled mind the nature of these beautiful and wondrous crystalline bodies, which no man has yet been able to fully comprehend and explain, without becoming aware of a definite intent on the part of the Creator.

Chapter I

HISTORY OF THE TOURMALINE. — ITS DISCOVERY, AND THE DERIVATION OF ITS NAME.

The tourmaline, even at the present day, is but little known, except to the amateur or the mineralogist; yet it is, perhaps, the most interesting of all the gems, when we come to consider the beauty and diversity of its color, the complexity of its composition, and the wonders of its physical properties. Although it has been exposed in the gem marts of Europe for a long time, yet its identity and true description are hardly a century old. The gem must have been known to the ancients, although there is no direct allusion to it by the gem-writers of the early periods. However, it is more than probable that some of the red tourmalines of the Uralian and Altian range of mountains in Siberia were gathered with those beautiful sky-blue beryls which were so highly prized at Rome, and which were then brought from the distant wilds of Scythia; and when the victory of Pompey over Mithridates fully made known to the Romans the fascinating beauty of Oriental gems, and awakened a taste for pearls, porcelain, and precious stones, the tourmaline must have been found among the varieties of gems that were brought from the gem mines of Ceylon and India.

The regular commerce established by the Arabs between these countries for a long period undoubtedly introduced into the bazaars of the Mediterranean some of these stones. But their characteristics were overlooked; or they may have been confounded with other minerals, the precise nature of which was then but imperfectly understood. The earliest description among ancient writers the antiquary can discover as applicable to the stone in question is in the writings of the Greek philosopher Theophrastus, who mentions a stone found in the Island of Cyprus, and which exhibited the unusual appearance of being emerald-green at one end, while it was of jasper-red at the other. This description may be readily applied to the tourmaline; for its crystals are sometimes red at one extremity, and green at the other: and, moreover, this peculiarity in the distribution of color belongs to no other gem. Crystals of sapphire are sometimes red and blue at their extremities, or yellow and blue; but a specimen displaying red and green hues in the same crystal is unknown.

Similar crystals of tourmaline have been found on another island of the Mediterranean Sea; and there is now, in a cabinet of minerals in this country, a tourmaline of light red and green which was found in the iron mines of Elba. Pliny, three centuries later, speaks in a vague manner of violet and brown stones, which acquired the curious property of attracting light bodies when heated in the rays of the sun, or even when warmed by friction with the fingers. In this characteristic description the mineralogist will at once recognize the tourmaline, and not the sapphire, nor the brown and violet topaz; for they were then unknown. Had the Latin philosopher intended to describe a peculiarity of the sapphire, he would undoubtedly have mentioned the red, the blue, and the yellow varieties also; for they all exhibit the same electrical properties alike, but not in so marked a degree as the tourmaline. Tourmalines of these particular colors are found at the present day in the mines of Ceylon and Pegu, and in the same districts which have yielded precious stones to commerce for more than two thousand years.

Beyond these obscure remarks, history does not devote a line nor a word to this remarkable stone; and for seventeen hundred years it is completely overlooked or forgotten.

In the middle of the seventeenth century, Brazil sent to the European market, among her exports, long prisms of dark-green stone; but De Laet, who wrote on gems at that time, and who ought to have recognized the crystals by their well-marked characteristics, simply termed them Brazilian emeralds, and also incorrectly asserted that they were harder than the true emerald of Peru.

At length the discovery came, and in the simple way that many of the truths of Nature have been revealed. On a warm summer's day early in the eighteenth century, some children were at play in a court-yard at Amsterdam. Their playthings lay exposed to the rays of the sun on the hot pavement; and among them were some of the precious stones the Dutch navigators had brought home from the gem-producing countries in the East Indies, and especially from Ceylon, which was then in possession of Holland. The children were astonished to behold some of the stones suddenly possessed with a strange power; for they attracted or repelled with a decided force ashes, straws, and other substances of little weight. The parents were summoned to view the strange scene.

The stolid Dutch lapidaries were, however, sorely perplexed at the mysterious action of the stones, which seemed endowed with the principle of life or motion; but, totally unable to determine the nature of the stone, or explain its singular properties, they gravely, and perhaps wisely, termed them *aschentreckers*, or "ash-drawers."

This curious story having been circulated abroad, it came to the hearing of M. Lemery, a gentleman of scientific tastes; and, after procuring a specimen of the stone, he exhibited its powers of attraction and repulsion to the French Academy of Sciences in 1727. But here all investigation and scientific inquiry ceased for nearly forty years. At this time a German physician, by name Æpinus, became interested in the report concerning the strange properties of the stone; and, obtaining two from Mr. Lechman, he commenced a series of experiments connected with the effect of heat and friction. After making a careful investigation of the subject, he published the results of his inquiries and experiments in the History of the Academy of Sciences at Berlin in 1756. The scientific world was startled at last with the statements of his memoir; and experimentalists over all Europe made haste to obtain specimens of the wonderful stone. The Duke de Noya, an Italian nobleman, obtaining two from Holland, hastily made some experiments in a careless way, and submitted his report to the French Academy. The duke, led astray by the imperfections of his observations, objected to some of the statements of Æpinus, and openly pronounced them incorrect. At this juncture of affairs, Dufay, Haller, Adamson, Colomb, and other *savans*, came to the rescue of Æpinus, and proved by their careful and varied experiments that he was correct. In England the spirit of inquiry was also aroused: and Dr. Heberden, obtaining the only tourmaline which was then in Great Britain, called to his aid some of the English philosophers; and they also confirmed the received opinion of its wonderful electric power. Fashionable society also became interested in the excited discussions of the philosophers; and the disputed stones were eagerly sought for by the fop as well as by the philosopher or the *dilettante.*

One of Hogarth's pictures, painted at this period, represents a gay youth arrested while absorbed with the wonders of the tourmaline when held up to the rays of the sun. Mr. Wilson and Mr. Carlton, two of the most eminent electricians of their day, also procured a number of tourmalines from Holland, and submitted them to numerous and interesting experiments, which fully sustained the assertions of Æpinus. Dr. Franklin became interested in the subject, and, after experimenting upon one of the stones, wrote to Dr. Heberden in 1759, supporting his theories. During the hot discussion which prevailed, the stones were described under various names, and were by some geologists supposed to be identical with the "lyncurium" of the ancients, which is now known to be the yellow zircon.

As all the accessible specimens had been cut by the Dutch lapidaries, or were in rounded masses like water-worn pebbles, the form of crystallization, and the common characteristics of tourmalines and schorl, were not then recognized. Linnaeus, in 1768, was the first to intimate their relationship; but it was reserved for Romè De Lisle to describe the Ceylon crystals, and establish their identity with the black variety, which had been known in Europe for almost two centuries. It was, however, a long time before the disputants adopted the present name of *tourmaline,* which is derived from the ancient Cingalese word *turmali,* and not *turamali,* which we understand is applied by the natives of Ceylon to the zircon. This remarkable mineral belongs evidently to rocks of crystallization, and never appears in the secondary terrains, or rocks of transition, with the exception of the gem mines of Ceylon, and perhaps Burmah. It is found generally in granite, mica schist, talc or talcose schist; and is divided by the chemist Rammelsberg into five sub-groups; viz., the magnesia, the iron magnesia, the iron, the iron manganese, and lithia and the lithia tourmalines. It is to the fourth and fifth group that our memoir is especially directed.

Tourmalines of these two groups often occur in beautifully-crystallized three-sided prisms, terminated by three principal planes, which sometimes are set on one extremity of one of the sides of the prism, and, on the other, on the edges. Its primitive form of crystallization is the obtuse rhomboid, having the axis parallel to the axis of the prism. The edges of these prisms are often truncated; and then the crystals form prisms of nine or twelve sides. However, it sometimes occurs massive and compact, or in parallel, divergent, radiating, and detached crystals. Its fracture is decidedly conchoidal, exhibiting internally a vitreous lustre. Its specific gravity ranges from 3 to 3.3; and its refractive power is 1.66, being superior to the topaz in brilliancy. Its hardness is 7.5, and quite equal to that of the emerald.

The tourmaline has as great a variety of names and synonymes as the sapphire; and in both minerals they arise from the great diversity of colors displayed by them. The red variety is known among mineralogists as the rubellite, siberite, or daourite; the blue as the indicolite; the white as achroite; and the black as aphrizite, or schorl. But, at the present day, they are all grouped under one name.

CHAPTER II

LOCALITIES AND PECULIARITIES OF DEPOSITION AND DISTRIBUTION

The principal localities for the transparent and finest specimens of the tourmaline are in Siberia, Brazil, Ceylon, and the State of Maine of the United States. In Siberia they are found in masses of felspar and quartz in coarse granite. There are several localities in this great country,—some near Ekaterinsburg and Sarapulsk, and others at Nertschink in Eastern Siberia, and near the confines of Northern China. The tourmalines found at these places exhibit a great variety of color: among them occur stones of the true ruby tint (the pigeon's-blood hue), and various shades of purple and of green. The arrangement of color observed in some of the crystals is quite remarkable, and differs from that seen in the specimens from other parts of the world. Some of these stones exhibit internally a blue or brown color, surrounded on the outside with a bright carmine-red, or of a dull yellow, waxy hue. Others may be red internally, enveloped with a pistachio green. Sometimes the crystals are of huge dimensions, and built up of solid material or of a multitude of acicular crystals; or they may occur in long, needle-like forms.

A great variety of color is displayed in these crystals. They may be pink at the summit, and light green at the base; or crimson, tipped with black iron ore. They may also be of a resinous yellow, coated with carmine of intense hue; or of a dark green, changing into an indigo-bluish tinge. The finest red-tinted tourmalines yet known have come from some of these localities. Some of these stones resemble the red sapphire known as the Oriental ruby so perfectly in color, that it is impossible to distinguish them by the eye alone. They are extremely rare, and are as eagerly sought for by the *dilettante* at the same enormous price of the true ruby. We shall not be surprised if the magnificent ruby in the Russian crown of the Empress Anne Ivanovna proves to be a tourmaline; and it will not fall in our estimation of its value if it is a siberite. It came from Pekin, which is not far distant from the tourmaline mines of Nertschink. Of the same nature may be the monster red gem which hangs as a pendant to the jade necklace which belonged to the Chinese emperor, and which was captured by the French in the sack of Pekin. This variety of the tourmaline is very apt to be flawed, or filled with imperfections, and especially with hollow threads and feathers and fibres, which are rarely seen in the green, the yellow, or the blue varieties. It is certainly curious that this variety should be so much more liable to imperfections than any of the others, not excepting even the purple.

Purple, blue, and green varieties come from Brazil; but, concerning the formation in which they occur, we can learn but little. We judge, from the famous dispute in the seventeenth century among the Jesuits about a mine at Esperitu, that it is mined in Brazil as well as in Siberia. The crystals brought to us from that country generally do not show signs of having been rolled in the drift; for the striæ of their sides are perfect. The absence and loss of perfectly faceted summits do not prove external violence, but rather a submission to the action of the elements, as observed in other localities. They occur of various shades of green, from the light tint to the darkest bottle-green. Sometimes we find them of a beautiful Berlin blue, or of a crimson passing insensibly into a bluish white or a bluish green. The most beautiful specimen yet produced by this country was not long since in the cabinet of minerals of the Duke of Florence.

It exhibited five splendid crystals of dark-green color on a matrix almost a foot square. Three of these superb crystals were erect, and one prostrate: they were fine prisms, and measured from two to four inches in length by three-quarters of an inch to an inch in diameter.

In Ceylon—the land of gems—the tourmaline, with the exception of the black variety, has not yet been discovered in place: they are always found there in the same gravel-beds of secondary formation with the sapphire, among the *débris* of the rocks of crystallization. Often they occur in rolled masses or in natural nodules, and sometimes in crystals whose faces are uninjured, and whose angles are unbroken. The question may arise, How can this mineral occur in nodules, when the laws relating to its deposition and crystallization are apparently so rigid? We will not attempt to solve the mystery, but can produce many examples to prove that the tourmaline may deposit a nodular concretion in the midst of a perfect crystal. We have removed many tourmalines from the cavities at the locality in Maine which exhibit this peculiarity: the crystals were of perfect form, but shattered by the elements; and, as we attempted to remove them from their beds in the sand, the sharp angles and striated sides fell into minute fragments, leaving a bright rounded nodule in the midst of the stone. Sometimes this nodule would be near the summit or the base, and sometimes in the central portion.

If Nature can deposit the tourmaline in this form in the midst of well-defined crystallization, we can see no objection to the same deposition without the more perfect combinations. We may also observe this tendency of Nature to globular concretion in several other minerals. It is well marked in the sapphires found in Ceylon in the same deposits with the tourmalines.

We may also find it illustrated in the diamond, and particularly in the white topaz and the chrysoberyl. The lapidaries notice that these nodules are more difficult to cut than are the well-defined crystals of the same mineral; and they also observe that they exhibit greater brilliancy. It has been well demonstrated that a perfectly formed and limpid crystal will not cut so brilliant a gem as one of these singular formations. In a future article upon the sapphire we will describe at length the peculiarities of the occurrence of the stone in the secondary formations of Ceylon, and attempt to explain some of the circumstances attending the manner of their deposition. The gem mines of Ceylon have yielded these stones for several thousand years, and probably supplied India and Persia for a long period of time. The range of color of the tourmalines found there is not very extensive: we have seen them of an impure green, and of various shades of yellow and brown; we have also evidence of red, gray, and hyacinth varieties having been obtained from the same locality.

Ava, that unknown land of rubies, also occasionally sends forth into the commercial world some beautiful specimens of this gem; but every thing that relates to the mines—their locality, or the quantity of gems exhumed—is shrouded in darkness and mystery. The Burmese government refuse all access to the foreign explorer, and restricts the export of all her mineral treasures; but when the British embassy, under Col. Symes, visited the Burmese empire in 1795, the mogul presented them, among other gifts, a magnificent group of pink tourmalines. It is composed of several crystals beautifully striated and terminated, arising from a matrix of what appears to be amethystine quartz.

The specimen has been valued at one thousand pounds sterling, and is now placed in the British Museum. In this priceless depository of art and the natural treasures of the earth the display of tourmalines is grand, and of the value of many thousand dollars. Every locality known on the globe is represented; and every variety of the stone, with its associate minerals, is exhibited. Among the rare specimens are some monster pink and crimson crystals from Ava, one of the kingdoms of the Burmese empire: these are composed of a multitude of long acicular crystals, sometimes rigidly straight and sometimes curved, and reminding one of miniature columns of basalt. This same peculiar arrangement of crystalline grouping is also witnessed in several of the tourmalines from Elba and Siberia, but not with so marked distinctness, nor on so grand a scale. At Roschnia in Moravia, it occurs in rose-colored and violet varieties in quartz and lepidolite in gneiss. Uto in Sweden yields fine crystals of deep red, purple, and blue, in a matrix of granite. The group of mountains of which St. Gothard is composed afford clear light-green specimens in dolomite, and which resemble in color the beautiful green beryls from the Ural Mountains. At Ariolo, at the foot of the St. Gothard pass on the Italian side, the rare white variety is found, but not in much perfection. The Island of Elba produces crystals of tourmaline of various colors, associated with the famous iron ore, and deposited in steatite. Some of these crystals exhibit the tendency to acicular grouping which is so well marked in the specimens from Siberia and Ava. The colors of these tourmalines are not often of deep and decided hues, but pale in their tints. The rare white variety, known as the "achroite," is found here in greater abundance than in any other known locality: they are, however, generally translucent, or too defective to be cut into gems. We have seen specimens from this island which exhibit transparent white color at one extremity, and a light grass-green with a yellow tinge at the other; or of a light yellowish green throughout, tipped at both ends with iron ore. Pink is the most frequent color met with at this locality: although we have known them of a light green, changing at one extremity into quite an olive-black; or of a light rose-pink at one end, and pea-green at the other.

The mountains of the Tyrol are famous for their brown tourmalines, which excel all those of other known localities. The best deposit was discovered by Mr. Wilkes. This traveller found them in a vein of talc and steatite situated in granite, on the lofty mountain known as Mt. Grenier, at the extremity of the valley of Zillethal. These tourmalines, which are called "tourmalines of Carenthia," are of a light yellowish brown, and sometimes of darker shades. They often occur in fine crystals, with both terminations faceted alike; which is an exception to the law of crystallography. Brown tourmalines of similar hue are also found in highly modified crystals at Gouverneur, N.Y., in a bed of granular limestone.

There are many localities in America where this variety occurs in more or less perfection; but the finest specimens come from Canada. Grand Calumet Island yields superb crystals of greenish yellow an inch thick; and Burgess produces specimens of a rich golden brown, reminding the amateur of that transcendent gem, the orange-tinted jargoon of Ceylon.

In the United States, there are but few localities where the colored tourmalines occur; and the best are found in the States of Massachusetts and Maine. In the town of Chesterfield, in the State of Massachusetts, is a noted locality of green, red, and blue tourmalines. They are contained in granite which is crossed obliquely by a vein of smoky quartz and a silicious felspar known as "Cleavelandite." This vein is of a width varying from six to eighteen inches, and contains the tourmaline crystals. These tourmalines are embedded in the quartz and felspar, sometimes passing through both. They are in long rounded prisms, deeply striated longitudinally; and often exhibit well-defined, trihedral summits. They have been observed four inches long, and more than an inch in diameter; but the solitary red tourmalines are rarely ever over one-fourth of an inch in diameter. Some of the crystals are dark green; others pink, red, violet, or dark blue, and sometimes of a light blue. They are nearly always quite opaque; sometimes translucent, but never sufficiently transparent to serve as gems.

The arrangement of the crystals sometimes presents a singular appearance. Well-defined crystals of rubellite may be seen completely enclosed in a crystal of dark-green tourmaline. In one instance, three red crystals were aggregated together, and enclosed by one of green. The green crystals sometimes embrace indicolite, and are often bent and dislocated by some unknown force. The energy of the dislocation, in several instances, has separated portions of a crystal, and subjected them to a kind of *échelon* movement, indicating that the disturbance took place before the crystallization was complete. Some of the crystals which contain the red tourmalines within the green are very remarkable; for they seem to be quite distinct, although the sides and angles of both prisms correspond. Sometimes, however, the separation is well marked by a thin layer of talc intervening between the two prisms. In some parts of the vein of smoky quartz, nearly every green tourmaline was observed to contain a rubellite. At the town of Goshen, six miles distant from the Chesterfield locality, the same formation occurs, but with a change in the colors of the tourmalines.

Here the red is rare; while the green, light azure, and dark blue are more abundant. The blue sometimes passes insensibly into green, becoming quite transparent. We have seen one crystal of red, tinged with blue, that might be considered a gem. A few crystals have been observed of a brown hue, changing to nearly white; and others have been seen of a yellowish green. They occur not only in well-defined prisms, but are often seen of acicular forms, and in radiated groups. Both of these deposits are now said to be exhausted, having been quite superficial in their extent.

CRYSTAL OF TOURMALINE
Exact Size.

Hamlin Collection.   Mt. Mica, ME

Chapter III

Description of Mt. Mica and Its Mineral Treasures

The most remarkable locality of the tourmaline in the United States, and which is also one of the most celebrated in the world, is in the town of Paris, in the State of Maine. It occurs on the brow of a little hill, which has been named by the mineralogists Mt. Mica, from the abundance of the muscovite which occurs there. The hill is one of the spurs of a more considerable elevation called Streaked Mountain, from the rugged and denuded appearance of its sides. It is but few rods square in extent, and is covered with turf and alluvial earth, with the exception of a little space in the centre and at the summit, where the ledge bursts out to the view. Although it appears coarse and utterly valueless to the casual glance, it is, nevertheless, one of the most remarkable mineral deposits on the face of the globe; for it has yielded from an area thirty feet square nearly forty varieties of minerals, some of them of extreme beauty and rarity.

It was discovered in the year 1820 by two students by the name of Elijah L. Hamlin and Ezekiel Holmes. They had been searching for minerals during the day along the mountain-ridge to the southward, and were then descending the western declivity on their way to the village. It was on the last day of autumn; and the glimmering rays of the setting sun were gilding with renewed splendor the faded colors of the landscape as the students were passing over the top of one of the lowest knolls. The view of the distant mountains (which are the loftiest in New England), the intervening valleys softened with purple shadows, the patches of green grass in the meadows untouched by early frost, the variegated hues of the forest- leaves left by the autumnal winds, the broad extents of russet brown of the stubble-fields, contrasting vividly with the glorious hues of the sunset sky, composed a scene of exquisite loveliness. The youths, spell-bound by the entrancing beauty of the landscape, lingered upon the hill-top until the valleys were shrouded with the shadows of commencing twilight. As they turned to descend the hillock, a vivid gleam of green flashed from an object on the roots of a tree upturned by the wind, and caught the eye of young Hamlin. Advancing to the spot, he perceived a fragment of a transparent green crystal lying loose upon some earth which still clung to the root of the fallen tree. The student clutched the gem with eagerness; and calling back his companion, who had passed over the brow of the hill, they closely searched the surrounding soil for other specimens. But the rapidly-increasing twilight soon compelled the youthful mineralogists to abandon the search. They, however, resolved to return at daybreak, and continue the exploration. But during the night a storm arose, and covered the hill and its adjacent fields with a thick mantle of snow, which remained until spring.

As soon as the winter snows had melted away, and left the hill and its sides exposed, the students returned to the search. They went directly to the ledge, which crops out on the summit of the hill, and which they had not examined before darkness overtook them on their previous visit. As they climbed up over the smooth and denuded surface of the rock, they were astonished to observe many crystals, and fragments of crystals, lying exposed upon the bare ledge, and sparkling in the rays of the sun. These they carefully gathered; and tracing others to the earth below the ledge, and which had formed from the decomposition of the rock, they eagerly turned up the soil in search of its hidden treasures. Thirty or more crystals of remarkable transparency and beauty rewarded the labors of the students; and with joy they held them up to the sunlight, and admired their varied colors of green, red, white, and yellow, of different shades.

They had, indeed, stumbled upon one of the richest and rarest of Nature's laboratories. All around the brow of the ledge, enormous masses of rose-red lepidolite, splendid groups of crystallized quartz of white and smoky hues, crystals of tin, broad foliæ of glistening mica, snowy flakes of felspar, studded with transparent green and red tourmalines, lay scattered about in profusion. Collecting as many of the choice and beautiful specimens as they could carry, the students, heavily laden, returned to the village, and sought to ascertain the nature of their mineral treasures. Subsequent examination indicated that the ledge was perforated with cavities, in which the tourmalines and other rare minerals had been deposited. It was also evident that the crystals that had been gathered up by the students had been set free from their cavities by the decomposition of indefinite periods of time, which had removed the surface of the ledge. There was no evidence of drift; and the crystals lay exposed upon the rock; while the softer materials had been washed by the rain down to the base of the ledge, and accumulated as soil. Parts of the ledge yet exposed to view were fairly honeycombed with small cavities and soft spots, where the decomposing felspar was crumbling away. In these cavities and decayed places in the rock other tourmalines were obtained by breaking away the edges of the ledge, or removing the decomposed stone.

The discovery having been made known to the villagers, many of them hastened to the spot, and secured a number of fine specimens as trophies or mementoes. As no one in the vicinity was able to distinguish the character of the gems, or even make known their name, the students enclosed a few of the smaller crystals in a letter to Prof. Silliman, and requested him to describe them. He kindly and promptly informed the youths that the minerals were tourmalines, and of rare occurrence. Thereupon the students selected some of the finest and purest of the crystals, and addressed them to the professor in return for his kindness. The parcel was intrusted for safe keeping to the late Gov. Lincoln, who was then a member of Congress, and about to start for Washington. At this period the journey to the capital was a serious undertaking; and the condition of the roads required that it should be made on horseback, at least for a great part of the distance. The governor started safely with the precious package, but lost it before reaching New Haven; and no trace of it has ever been found.

Two years after the discovery, the younger brothers of the discoverer, Cyrus and Hannibal Hamlin, although scarcely in their teens, resolved to make an attempt at a more complete exploration of the ledge. Having borrowed some blasting-tools in the village, they proceeded to the hill, and managed, in a rough way, to drill four or five holes in the surface of the ledge, and blast them out. These operations, though of trivial magnitude, were attended with unlooked-for success; for the explosions threw out, to the astonishment of the boys, large quantities of bright-colored lepidolite, broad foliæ of transparent mica, and masses of quartz crystals of a variety of hues. The last blast exposed a decayed spot in the ledge, which yielded readily to the thrusts of a sharpened stick or the point of the iron drills. As the surface was removed, great numbers of minute tourmalines were discovered in the decomposed felspar and lepidolite. The rock became softer and softer as the boys proceeded in their labor of excavation; and soon they reached a large cavity of two or more bushels' capacity. This cavity, which was situated in the heart of the solid ledge, was filled with a substance which appeared to be sand, loosely packed. Amongst this sand, or disintegrated rock, crystals of tourmaline of extraordinary beauty were found scattered here and there in the soft matrix. Scratching away with renewed energy, the boys soon emptied the pocket of its contents, and found that they had obtained more than twenty splendid crystals of various forms and hues. One of these was a magnificent tourmaline of a rich green color and remarkable transparency. It was more than two inches and a half in length by nearly two inches in diameter; and both of its terminations were finely formed, and were perfect. Several others possessed extraordinary beauty; and some of them were fully three inches in length, and an inch in diameter. The colors of these tourmalines were quite varied, but were chiefly red and green, and far surpassed in the purity and transparency of their hues the crystals collected by Elijah Hamlin in his previous examination of the locality. The exact number of crystals obtained is not now known; but when collected together, with the fragments of others, they filled a basket of nearly two quarts' capacity. Besides the tourmalines, the quantity of lepidolite, mica, and other choice minerals, thrown out by the blasts, or found in the sides of the cavity, was so great, that the boys were obliged to seek for an ox-team to transport them home. So little was known of the value of these rare minerals at that time, that the possessors considered the finest of their treasures to be worth about a guinea. Cyrus had learned from his brother Elijah, who was then living in the eastern part of the State, the names of some European mineralogists who had made inquiries of him concerning the discovery of Mt. Mica and the disposition of its minerals. With some of these he placed himself in communication, and from time to time disposed of nearly all of the finest of the crystals in exchange for money or minerals. Cyrus afterwards moved to Texas, where he died many years ago; and with him has perished the history and distribution of these gems.

The younger brother and survivor, Hannibal, took but little interest in mineralogy, and gave his share to his brother. He now remembers only the facts of the discovery, the curious and symmetrical forms, the perfect limpidity, and the wonderful beauty, of the crystals. This is all that is known of the history of the splendid gems and wonderful crystals that Mt. Mica yielded to the explorer in its early and best days. Gathered then in profusion, and carelessly treasured, they have since been scattered over the world, and, in many instances, their identity lost. The late Prof. Cleaveland, a famous mineralogist in his day, received several fine crystals, and among them a superb yellow tourmaline of the purest water. There is now no trace left of these specimens. His cabinet, which Bowdoin College inherited, does not now contain them; but, from the evidence gleaned from his correspondence, it is surmised that they may have been sent to his friend, the celebrated Berzelius, and are now in the mineralogical cabinets of Sweden. Those that fell to the share of young Holmes at the time of the discovery were destroyed many years ago in the fire that burnt the Gardiner Lyceum. In the Imperial Collection of Minerals at Vienna, there are some tourmalines of remarkable beauty; and mineralogists are at once struck with their perfect resemblance to the Maine tourmalines, especially in their arrangement of color. They came from the cabinet of the antiquary Vander Null, and were simply labelled "America." This is all that is definitely known of them. As the tourmalines of all known localities have peculiarities which distinguish them in a marked decree from each other when viewed by the practised eye, it is easy for the mineralogist to give the locality to the unlettered specimen. Moreover, from the evidence now in our possession, we feel confident that these tourmalines at Vienna are a part of the results of the early exploration of Mt. Mica. Baron Lœderer, an experienced mineralogist, happened to be in Vienna when the Austrian government purchased the collection of Vander Null. He was present at the museum when the boxes were opened; and as he had visited Mt. Mica previously, and was familiar with the peculiarities of the mineral, he at once recognized them to be identical with the tourmalines of

Maine. He believed them to have been taken from Mt. Mica previous to the year 1825. The baron is now dead; but this information was communicated to a geologist in this country prior to 1830. In the Vienna cabinet there is one crystal of tourmaline with both terminations complete. Among those found by the Hamlin boys was a crystal that coincides with this description; and tourmalines of this perfection are of extraordinary rarity.

In 1825, five years after the discovery, Prof. Shepard, a young and enthusiastic mineralogist, visited the locality, and observed a decayed place in the ledge, where a mass of felspar had become decomposed. By digging out this substance, and removing the superincumbent earth, a drusy cavity, or series of cavities, three feet in length and two in depth, were exposed to the gaze of the fortunate explorer. At the bottom of these cavities, among particles of cookeite, lepidolite, and other decomposed minerals, resembling sand, lay a number of magnificent tourmalines of perfect transparency, and exhibiting colors of red, blue, green, and also of variegated hues. Some of these splendid crystals were several inches in length, and more than an inch in diameter; but, unfortunately, they were not in perfect condition. The rain trickling down from the surface of the ledge, through its crevices, had, by the effects of freezing and thawing, cracked and shattered portions of the crystals. The terminations of some of them were broken into fragments; while the shafts remained entire, or slightly fissured. Some of these prisms were green at one extremity, and ruby-red at the other, or green on the exterior, changing imperceptibly to a beautiful crimson in the interior. Others were entirely green or red or blue or variegated.

The fame of Mt. Mica now became known far and wide; and mineralogists from all parts of the country hastened to visit and explore the locality. The Russian and Austrian consuls, Mr. Cramer and Baron Lœderer, both enthusiastic collectors, examined the deposit, and carried away large quantities of fine specimens.

All the accessible part of the ledge had now been explored; and mining operations of a more solid character were required to follow the continuation of the deposit, which still appeared at the bottom of the excavations. The ill-fated Prof. Webster blasted down a few feet, and opened a cavity which yielded a grass-green crystal of great purity, quite as long as the finger. At a subsequent time he discovered another cavity, from which he drew out a superb red crystal the size of the thumb. The excited and overjoyed professor sprang to the top of the ledge, and, holding up the beautiful gem in the rays of the sunlight, danced over the rock like a madman, exclaiming that he would not take a large sum of money for it. Nothing more is known concerning these remarkable specimens; but it is surmised that they were sent to Europe, and were probably cut into gems, and may now adorn some of the royal crowns.

From time to time, during a period of more than forty years, many other explorers visited the locality; but they examined it in a superficial manner; and in the year 1865 the deposit was regarded by mineralogists as completely exhausted, although the excavation in the ledge did not exceed fifteen feet square, nor more than six feet in depth. At this time the writer carefully examined the hill, and found no signs of tourmalines, with the exception of a small piece of lepidolite, which appeared in the pit at the base of its southern wall. With the aid of a miner we placed a blast in the rear of the lepidolite; and, to our joy, the explosion revealed a small cavity about the size of the fist, in which lay a crystal of green tourmaline tipped with red, and an inch in length. Encouraged by this success, we commenced a series of careful explorations, which, undertaken at various times extended over a period of three years.

During this reconnoissance, we removed an extent of ledge averaging about six feet in depth, and amounting in all to more than one hundred tons. Three cavities only were exposed by these explorations; and, as no sign of the mineral deposit remained in sight to cheer the explorer, all further research was then abandoned. All of these three cavities were situated at a depth of six feet from the surface, and contained fragments and *débris* of what were once beautiful crystals of tourmaline. But the water and the action of the frost had, even at this depth, exerted their mighty force upon the mineral, and had rent their solid and transparent forms into numberless fragments. The crystals lay in their sandy beds undisturbed in regularity of outline; but they crumbled away as soon as touched. Here a summit of crystal with faceted planes would be preserved, while the rest was destroyed; and there the base or a nodule from the central portion would alone remain among the wreck of the marvel of Nature's work. The base and sides of these cavities were composed of quartz mixed with lepidolite and other firm minerals, forming natural basins, into which the water trickled down from the ledge above through its numberless crevices; and so the tourmalines were constantly exposed to the action of water, until the walls of the cavities became rent, and the water allowed to escape to deeper outlets.

The year following this abandonment of the mine, a party of explorers, searching for mica for commercial purposes, commenced operations at the same place, and proceeded to remove the rock on the eastern side of the pit. They removed about three hundred tons of rock, and descended to the depth of quite eight feet. At nearly this depth, the miners struck five well-defined cavities on a line ranging from east to west, but disconnected with each other. All these cavities contained tourmalines in broken crystals of various colors; and in one of them was deposited in and on a mass of white quartz one of the most remarkable groups of tourmalines yet discovered in any part of the world. Separated into fragments by the ignorant miners, they were scattered in various cabinets, and some even cut into gems, before the mass of quartz which served as the matrix was discovered. However, their dimensions were preserved; and from the remaining crystals and fragments the group has been reconstructed in miniature. The mass of quartz was about eight inches square, and five in depth. On its summit arose a crystal of tourmaline two inches in diameter, and fully two and a half in height. It was transparent; pink at its base; changing, towards the summit, to a delicate and gorgeous carmine of considerable depth of hue. On the side of the quartz matrix appeared a fine prism fully three inches in length, and three-fourths of an inch in its longest diameter. This crystal was transparent, and of the purest grass-green; in fact, some of its fragments cut gems resembling very closely the finest of the Peruvian emeralds. Another crystal, of unknown length, but more than an inch in diameter, was of a beautiful blue-green in its centre, surrounded with a coating of clear white tourmaline a line in depth. This was also surrounded by three other layers of transparent tourmaline, each about a line in depth. The first was pink in hue; the next, limpid white; the last, and the exterior, was a soft celandine green. The fragment which has been preserved, when viewed axially, presents plainly this remarkable arrangement of color. There were other crystals of white and green, or white passing to a very light blue. The whole number of distinct crystals arising from the mass of quartz as a matrix were nine; and all were transparent.

The writer, again taking courage at the success of the mica-hunters, commenced explorations on the northern and eastern wall of the pit. Several fine specimens of rose-red lepidolite, and some other lithia minerals, appeared on the side of the excavation to give hope to the mineralogist. Eighty tons of rock were removed in this operation before a cavity was struck. One ton of lepidolite was obtained, including a large mass weighing five hundred pounds. The cavity proved to be a large one of more than a bushel capacity, and yielded a great number of minute crystals of tourmaline, besides several large specimens, which, unfortunately, were in a state of disintegration. Some months afterwards the exploration was continued, and in the same direction,—to the northeast. After removing forty tons of rock, a small cavity the size of the hand was opened, and yielded a broken crystal of dark green the size of the thumb, and a remarkably slender prism of bluish green more than three inches in length, and one-fourth of an inch in diameter. This singular specimen is a facsimile of some of the Siberian beryls, and will readily pass as such.

In this last exposure of the ledge, no lepidolite, and very few of the associate minerals that accompany the tourmalines, were obtained; and, from the appearance of the wall, the miners concluded that the eastern limit of the mineral deposit had been reached: therefore the exploration in that direction was stopped.

The next summer the western flank was examined; and, a few preliminary blasts having yielded positive signs, the miners were directed to blast out an extent of the ledge amounting to about sixty tons. During this removal, several decomposed spots in the albite, enclosing tourmalines, were discovered; and finally a large cavity was reached, which yielded many minute crystals of pure white tourmalines, and fragments of what were once magnificent crystals of white and red, and white and dark blue.

A month later in the season, the work of blasting out the western flank was resumed. Fifty tons of rock were removed; but not a single tourmaline, nor a specimen of the rare minerals associated with them, was obtained. We then arrived at the conclusion that both flanks of the deposit had been reached; and the only hope of obtaining further tourmalines lay in blasting out the central portion of the ledge. To reach the imaginary line of the tourmaline deposit will necessitate the removal of large quantities of rock to the depth of eight feet; and, as this operation will require a large expenditure of labor and money, all further attempts to explore the ledge have been abandoned. We do not, however, consider the locality as totally exhausted; but we regard all future mining operations in search of the tourmalines as extremely hazardous and costly.

From the data afforded by the removal of many hundred tons of rock, and the exposure of a large extent of the mineral-bearing portion of the ledge, we have arrived at the conclusion that there is no well-defined deposit; neither is there any semblance to a vein in which these minerals may be traced with a degree of certainty. But there is an indefinite arrangement, an imaginary line at a fixed depth below the surface, in which we find the tourmaline deposits, and, in fact, all the other rare minerals for which Mt. Mica has become famous. This line dips to the south-east, and descends gently from the place discovered by the students down to the bottom of the southeastern wall of the pit, where it disappears eight feet below the surface.

To describe this strange deposit in strictly scientific terms will indeed be a difficult task; but we will endeavor to make it appear in the same light to our readers as it appeared to us. The ledge, in its early days of examination, seemed to be foliated, but not stratified; and consisted of layers of granite, bending with gentle inclination toward the north-west. This inclination of the layers, at first gentle, is now observed, at the back of the pit, to be almost perpendicular. These folds of the granite lay like the leaves of a book, but not of a definite thickness. As they bent over to a certain extent, the coarse granite of the superior rocks suddenly changed in character. It was granite still; but the arrangement of its particles had undergone a decided change. The masses, flakes, and coarse crystals of albite, the large nodules of quartz, the broad plates of mica, and the huge and numerous crystals of schorl, vanished; and, instead of them, a ledge appeared of firmer texture, but composed of much smaller particles of the same materials. The line of demarcation was quite apparent; yet there was no line of decided and distinct separation. Along this imaginary streak of changed arrangement of material occurred the tourmaline deposits. They sometimes happened in the folds of granite a foot or two above this line, but never below it. Of all the twenty cavities known to us, we are not aware of a single one occurring below this change in the rock. The early explorers found the deposits at the surface, and followed them to the southward, about fifteen feet in distance, where the streak had declined to the depth of six feet below the surface. There was no direct communication between these cavities, or pockets; but the soft and partly-decomposed rock indicated deposits beneath or beyond. In fact, the fifteen feet square extent of ledge excavated in the early days was fairly honeycombed with cavities. But the later explorers were obliged to grope in the dark, and trust to hazard in their search for the mineral treasures. Cavities were suddenly found at a considerable distance from the first workings, and often when hope of success was nearly abandoned. The appearance of lepidolite was often a sign of coming success, especially when followed by masses of smoky quartz. When a broad layer of felspar was found to be changing into regular and broken flakes, a deposit or a cavity might safely be prophesied to occur beneath.

Interspersed throughout the ledge in great abundance appeared well-defined but shattered crystals of black tourmaline, some of them more than a foot in length. In the rear wall of the pit, a huge crystal nearly three feet in length may be seen to-day dislocated and shattered. But, strange to say, among all the cavities in which the transparent tourmalines were found, not a single crystal of the black variety occurred. It is a remarkable fact to be considered in the formation and deposition of this mineral.

The cavities generally were roofed with albite; whilst the sides were composed of limpid or smoky quartz mixed with lepidolite, crystals of tin, spodumene, amblygonite, and other rare minerals. These cavities were of irregular shapes, and of sizes extending from the capacity of a pint to that of two or more bushels. The interior was always filled with a substance resembling sand, but which is probably disintegrated cookeite and gray lepidolite. Lying loose in the sand, and generally at the bottom of the cavity, appeared the beautiful tourmalines, often unattached, and disconnected with any matrix except the loose sand. Sometimes, however, they were attached to the walls of the cavity, or, broken by unknown cause, became separated from their matrix. Occasionally the quartz rock would form fine crystals of pellucid or smoky quartz, which were often transfixed with slender crystals of tourmalines of various colors.

The walls of the cavities, though composed of the strongest materials, were often found rent and shattered by some unknown disturbing force; perhaps by electricity, but probably by the mighty effects of sudden contraction and expansion caused by the freezing of the water which trickled down through the crevices of the rock above, and exposed to the frosts of winter and the heats of summer. To these agencies do we feel inclined to ascribe the shattered condition of the crystals. Sometimes the shafts of the prisms were broken into two or three pieces; and in other instances they were fractured into numberless minute fragments. When the superincumbent sand was removed, the broken and disintegrated crystal might be seen in its bed with undisturbed outline; but, at the first touch, the symmetrical form crumbled into particles both coarse and minute. Nature had evidently constructed her forms of crystallization in absolute perfection; and the process of disintegration happened long afterwards, probably from external violence.

From the evidence collected by or known personally to us, we believe that Mt. Mica has yielded over a hundred crystals which would be considered as fine and remarkable specimens. Of the smaller tourmalines, ranging from one inch down to microscopic size, no fair estimate can be made; but they amount to many thousands. We have seen specimens containing more than fifty distinct and transparent crystals embedded in masses of lepidolite, cookeite, and albite. Coarse and opaque, or even translucent, crystals of tourmaline, several inches in diameter, and nearly a foot in length, have been found in the great masses of albite and quartz; but all the fine and transparent prisms have been taken from the cavities, with very few exceptions. These exceptions refer to a few crystals found in portions of felspar, which were soft and pliable, and of similar character to the distinct cavities.

From the data thus far obtained, it is also evident that this deposit which affords the tourmalines is of but little depth, and limited in its area. This superficial degree of deposit is not confined to the tourmaline alone, but is observed with most of the gems, and with some of the metals. It seems as though the light of heaven was required in the production of the gems, as it is for the marvellous and varied hues of the flowers of vegetation. Thus far, nearly all of our precious stones have been found on or near the surface of the earth; and it appears as though the contact of the air or a ray of sunlight

The Tourmaline

was required to build up their forms and perfect their hues. Down in the thousand mines along the slope of the Rocky Mountains the amethyst vanishes below the depth of twenty or thirty feet, while the same quartz crystallizes in its beautiful and definite but colorless forms in the depths of the deepest mines. The diamond and the sapphire belong to superficial terrains; and we find that the rule of shallow deposit relates to most of the gems. The topaz of Brazil, the beryl of Siberia, the chrysoprase of Silesia, the turquoise of Thibet, or the opals of Hungary, all occur near the surface of the earth, and are never found below a certain depth.

No other deposit in the world yet known to the mineralogist has yielded tourmalines of such a variety of color as Mt. Mica. Some of the fragments of the broken crystals rival in beauty and limpidity, even surpassing in brilliancy, the emeralds of Peru. Others are almost equal to the purest rubellites of Siberia, which resemble the red sapphire; or they imitate with a degree of perfection the dark-green crystals of Brazil, the light-green of St. Gothard, the pink of Elba, the light-yellow of Ceylon, the blue of Sweden, and the rare white of Ariolo.

The arrangement of color often observed in these minerals is very remarkable, and reminds one of the diverse coloring sometimes seen in the sapphire, but on a far more extended scale. In some of the crystals the red changes into blue, and the blue finally passes into green or black; or the red may pass into white, and the white be tipped with green. In others the color is simply red and green, or white and green, exhibiting many intermediate shades. Generally these transitions and gradations of color are imperceptible as they pass into each other.

But in some specimens the colors are not mingled in the least, and the line of demarcation is well defined and trenchant. So sharply distinct are these crystals in color, that they seem to be composed of several sections veneered together; yet these stones are homogeneous, and cannot be cleaved apart any more than the bands of the onyx.

One very beautiful crystal exhibits a most singular appearance of alterations; and its summit, which is regularly faceted with natural planes, is changed to white to the depth of a line. The contrast of this white cap to the green column of the crystal is so perfectly marked as to suggest the idea that it may be an accidental coating; but examination proves it to be an inseparable part of the crystal.

With the tourmalines of this locality we have noticed that the faceted terminations are always green; while the red is never seen except at the termination, which is flat; that is, in well-defined prisms. The crystals may appear entirely red; but they are not terminated nor well defined. This rule is not observed so markedly with the Siberian tourmalines; for with them some of the most beautifully faceted terminations are red. Sometimes the minute crystals may be found penetrating limpid quartz, like the specimens from Ekaterinsburg, which are cut into gems and ornamental stones. They then appear like arrows of rutile enclosed in quartz, but of red and green hues; and from their variety, as well as beautiful appearance, are highly prized by the Russians.

Masses of gray lepidolite and cookeite have been observed filled with crystals of tourmaline, hollow, like thin tubes of glass, with their interior coated completely or partially with yellow cookeite arranged in filaments, in tufts, or in masses. Some crystals have been found composed of a columnar structure, made up, as it were, with bundles of acicular crystals. Others have been observed strangely compressed in their form; and sometimes, when occurring in the mica, they have been reduced to a line in thickness, even when two inches in length.

Well-marked specimens of dislocated and curved crystals have frequently been found; and some beautifully-radiated tourmalines of a transparent green color have been exposed by rifting masses of mica. Sometimes we observe in the solid masses of quartz or felspar well-defined crystals of tourmalines articulated like pillars of basalt, and whose sections have been separated at some distance by the intervening rock. These singular modifications give rise to curious speculations as to their cause. The separation has evidently taken place while the crystal was forming; for the shaft of the prism is often complete and symmetrical, although its sections may be separated at the distance of several inches. This peculiarity is noticed with all the varieties, but is particularly marked in the black crystals: and so liable are they to this defect in homogeneity, that solid sections, or even masses, are rare; and a complete, unbroken prism has not yet been found.

What was the disturbing force? and at what period of the deposition of the mineral and its matrix did it take effect? These are themes of inquiry which are easy to conjecture, but difficult to determine.

Sometimes the acicular crystals are drawn out to a delicate fineness; and, in several instances, they have been seen arranged in groups, and as minute and silken as the thistle's down. Massive and opaque specimens have been occasionally met with; but they are rare: for the force of crystallization has left its impress upon almost every rock at Mt. Mica, and the rigidity of its laws has been well marked at this locality. Many perforated crystals have also been seen, occurring in thin, glass-like tubes sometimes more than an inch in length, but generally less. The interior of these singular crystals is often free from any substance; but some of them are filled with kaolin or cookeite of gray, white, or yellow shades.

In some of the masses of quartz, mixed with cookeite and lepidolite, remarkable but small cavities may be observed. These cavities are often empty; and their sides are beautifully striated, as though Nature had prepared a mould, and intended to deposit crystals of tourmalines therein, but had forgotten to do so. Some of these cavities are studded on their sides at random with minute transparent crystals of quartz partly covering the beautifully-defined striæ.

Some prisms have been observed transfixed by other crystals of tourmaline, indicating that opposite forces were at work during the process of crystallization. All the crystals have not perfect terminations; and often we meet them without any well-defined faces. Some of those found embedded in kaolin are of irregular form, and indicate that Nature, restrained by disturbing causes, has left her work imperfect both in symmetry and color. This hiatus in the regularity of the deposit is far more common in the pale pink tourmalines than in any other. Some of these irregular tourmalines are translucent, and composed of columnar-like masses, having the lustre and adamantine refraction peculiar to some varieties of felspar. Many of the crystals are coated in places with a thin deposition of cookeite, generally of a white, gray, or yellow hue. Some of the smaller prisms have been found completely enveloped with cookeite.

This rare mineral, cookeite, appears to be peculiar to the deposits at Mt. Mica and Hebron; for, amongst all the specimens of tourmaline we have seen from other parts of the world, not one exhibits the least trace of it. It belongs to the lithia group of minerals; and if it is really a product of alteration of the tourmaline, as some mineralogists suppose it to be, it is remarkable that it has not been found in some of the other tourmaline localities in Siberia or Brazil. Some large masses of cookeite mingled with gray lepidolite have been literally filled with transparent crystals of tourmalines of various colors, and of a great variety of shapes and conditions. Not only single prisms, but also multiple crystals and radiated forms, might be observed in these masses. Their deposition in the matrix, without any regular arrangement, recalls to mind the singular beryls of the Adun Tschilon in Siberia, and which lay at all angles in the mass of hydrous oxide of iron. Some of these remarkable cabinet specimens have been found containing fifty or more defined crystals of tourmaline. Transparent tourmalines of similar forms and colors have also been discovered in the side of an abrupt hill in the town of Hebron, about seven miles southeast of Mt. Mica. This locality was also discovered by accident. An itinerant lecturer who possessed a love for the beautiful and rare in Nature, and who had passed hours at Mt. Mica studying the geological formation of the place, was travelling through the town of Hebron, on his way to the neighboring village, not many years ago. As he rode along the valley, he espied a bowlder of rose-red lepidolite in the stone wall by the roadside. He recollected that this rock was one of the associate minerals of the tourmaline; and this discovery tempted him to seek for the parent ledge whence the mass had become detached. Climbing over the wall, he observed other fragments in the adjoining field; and, by tracing in this manner the surface of the ground, he soon came to the hillside, where the evidences were far more numerous, and where the deposit undoubtedly occurs. The ledge, however, is yet covered several feet deep with alluvial soil; and only the detached bowlders have been explored. These have yielded beautiful crystals of transparent and varied hues, and also the same association of minerals we observe at Mt. Mica. The surrounding earth has not been sifted, nor the ledge exposed: therefore it is not known whether the system of cavities prevails here as at Mt. Mica; but there is reason to presume it does, and that some bold explorer may one day reap a rich harvest for his labor.

Some curious specimens of altered rubellites have been found at this place. They seem to have changed into lepidolite, and still retain some of the characteristics of a natural crystal. As naturalists observe hybrids among animals, so mineralogists observe hybrids among minerals, resulting from mixture of isomorphous matters in all proportions. Sometimes one ingredient preponderates sufficiently to locate the mineral with the species to which it belongs; at other times they are all so evenly balanced, that it is difficult to determine the character of the compound: and so these rubellites may be found of all degrees of alteration, until their forms are lost, and the mineral is decidedly lepidolite.

It is a curious fact, that, by the agency of certain mysterious forces, pseudo-morphs are formed; that is, crystals undergo a change of composition without their forms being in the least degree affected. Even cavities are sometimes emptied of their contents, and minerals of a totally different character deposited therein. The locality at Hebron is now known as Mt. Rubellite, deriving its name from the abundance of red tourmaline.

Five years ago, the same strolling preacher called at the house of a physician in the town of Minot; and, observing on the mantle a specimen of rock containing a green tourmaline, he inquired whence it came. The doctor pointed to a naked ledge of rock on the brow of a hill in Auburn, nearly two miles distant, and informed the inquirer that it was the locality. Not at all daunted by the distance, our venerable enthusiast started for the hill, and soon made known to the world a new locality of transparent tourmalines. The name of the discoverer of these two localities is Luther Hills.

Early informed of the discovery, we called to the place our miners from Mt. Mica, and proceeded to explore the deposit. We found that the tourmalines appeared on the brow of a ledge which projected a little distance from a gentle slope of a hill, and far below its summit. The surface of the rock and adjoining earth was strewn with numerous foliæ of mica containing crystals of transparent tourmalines, and large masses of pink lepidolite, amounting in all to quite a ton in weight. The abundance of lepidolite and mica gave hope of an extensive deposit of the coveted crystals; and almost the first specimen picked up from the soil, exhibiting rich emerald-green hues, gave promise of superior gems. But we found, to our regret, that the deposit was very superficial, and was, in fact, a mere coating to the ledge.

A few blasts of the miners soon exposed the entire deposition of the tourmalines and their associate minerals.

The tourmalines found here were of the true emerald-green; and the specimen first found yielded a fine perfect gem of two karats, resembling perfectly the emeralds of Peru, but of a pale tint. Nearly all of the crystals were acicular, or in acicular groups; and some of them were beautifully radiated on plates of mica. Quite all of the specimens obtained were of emerald-green colors of various intensities of shade: very few exhibited faint pink or rose-red hues.

It is a little remarkable that all of these localities should occur on a direct line from each other, and invariably exposed to the west. The Auburn deposit occurs twelve miles south-east from Mt. Rubellite.

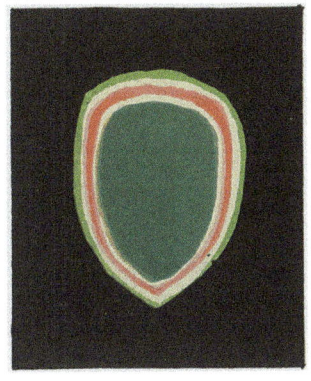

SECTION OF CRYSTAL OF TOURMALINE
Showing Arrangement of Colors.

Hamlin Collection.   Mt. Mica, ME

Chapter IV

The Physical Wonders of the Tourmaline—Its Electrical Properties—Its Optical Phenomena—Its Play of Color, etc., Compared with Other Gems

The tourmaline is exceedingly interesting to the student, on account of its complex mineralogical characters and curious physical properties; in which respect it far surpasses all the other gems. Its crystals are almost always differently terminated; which is an exception to the law of crystallography, that all facets of the same kind should similarly be reproduced on all identically similar elements of a crystal. It was thought, for a time, that this crystallographical anomaly exerted some relation over the electric properties of the stone; for, when one of its prisms was heated in a particular way, two kinds of electricity were manifested,—one end exhibiting positive electricity, and the other end becoming negative; and this had been observed so often as to be regarded as a uniform fact. But some of the rare tourmalines from Pegu and Ceylon, which possess terminations regular and exactly alike, likewise exhibit the same phenomena of the double property.

However, it generally manifests positive electricity at the termination which has the greatest number of facets; and this latent force may be easily aroused by friction or by the application of heat. This state of polarity may be reversed by intense cold;

and that which is positive suddenly becomes negative, and vice verse. If one of the prisms be broken while in an electric state, excited by heat, the fragments instantly present opposite poles, like artificial magnets. It is also shown, that, if it be heated somewhat above 212° Fahrenheit, it loses its electricity: but, if the increasing heat is continued to a certain degree, it again becomes excited; but its electrical poles are now reversed.

The proper degree of heat required for the exhibition of the electrical power of the tourmaline is from 100° Fahrenheit to 200°. Haüy discovered, that, by heating a crystal unequally in the focus of a lens or mirror, the position of the poles might be changed. To M. Æpinus, the German physician, and to Mr. Canton, the English electrician, is due most of the honor of making known the electrical properties; and especially to Mr. Canton, whose researches were published in the Proceedings of the Royal Society in 1759. Dr. Priestley, seven years later, turned his able attention to the peculiarities of the stone, and discovered a method of reversing all the experiments made upon it; making that side which is positive in heating or cooling to be negative, and that which is negative to be positive; so that the kind of electricity shall be just what the operator shall direct by the application of proper substances to the stone. These curious experiments will be found explained at length in the Philosophical Transactions, Franklin's Letters, and Priestley's Works. The phenomena of electricity exhibited by the gems are very interesting; and the nature of the subtle agent is quite as mysterious today as when the Milesian school of Thales, more than two thousand years ago, discovered the unknown force by the friction of a bit of amber. We find that some of them become electric with the greatest ease, like Iceland spar or the topaz; and sometimes they become excited by merely pressing them between the fingers. Others acquire the property with the greatest difficulty. Some retain the power for several days; while others, like the diamond and quartz, lose it almost instantly.

The transparent and perfect stones seem to possess more of this mysterious property than the translucent or opaque. The native garnet will not become electric by friction until its sides have been faceted by the lapidary. Generally all polished and transparent stones acquire positive electricity; but the effect is the reverse when they are rough like unpolished glass. The diamond is the exception to the rule, which relates to all combustibles which acquire negative electricity by friction. This property of acquiring electricity, and the comparative degree of strength, form important characters for determining the nature of gems, even after they have been polished, and set as jewels.

Some of the tourmalines exhibit the strange play of color which is called dichroism, and which is especially seen in perfection in the transparent mineral known as iolite. The term "dichroism" or "polychroism" is applied to a mineral when it displays two or more colors when viewed in different directions. The cause of this remarkable change of hue is still obscure, but is supposed to be due to a certain mixture of polarized with ordinary light, and is therefore only seen in minerals which possess double refraction. But few of the gems possess this singular property, even in a slight degree. The sapphire sometimes, but rarely, exhibits the play, but never in a decided manner, like the iolite. The rare and transparent andalusite from Brazil displays the property in a charming manner, and may exhibit in the same specimen delicate shades of white, green, and pink, according to the position in which the stone is viewed.

It is in the tourmaline, however, we witness the display of polychroism in its greatest perfection. Some of its crystals, when viewed parallel to their axes, appear of a splendid crimson hue; but, when the prism is slightly turned, the red color vanishes as if by magic, and the stone becomes white or smoky or green, without the least tinge of its former hue. Other crystals are green when viewed transversely, and yellowish brown axially; or they may be dark violet transversely, and greenish blue axially.

The range of diversity of color displayed by this mineral, when viewed in this manner, is very great; but all its crystals or masses do not possess the property with equal intensity. Some exhibit it with great distinctness; while other specimens display only a trace of it, and some none whatever. Turn the fragment whatever way you will, the color remains the same, and unchanged. This absence of dichroism is best observed in the light-colored specimens, which possess double refraction in a feeble degree.

The ancient glass-workers, and especially those of the third or fourth centuries, discovered the means of producing this strange optical effect at will; and there are still remaining some splendid examples of their skill. The two cups sent by Hadrian to his brother-in-law Servianus are very curious; for, although of a bright-green tint when seen by reflected light, they turn to a ruby-red when light is transmitted through them. The ancient glass relic in the collection of Baron Lionel Rothschild is of a pale ruby color by transmitted light, and of a pale opaque green by reflected light.

This phenomenon often appears under the influence of artificial light; and the laboratory of the chemist affords numerous and pleasing examples. Several of the compounds of chromium are green when seen by daylight, but change to a purplish red when viewed by candle-light. One of the most remarkable examples is shown by viewing a tree in full foliage, and holding between the eye and the object a flask containing a green solution of chromium. Although the tree and the solution may be of the same color, yet, when it is looked at through the solution, the foliage changes to a bright purplish red color.

The causes which give rise to the great variety of colors among the tourmalines, as well as the other gems, form interesting themes of inquiry; and they are yet subjects of controversy among scientific men, and probably always will be. The study of the origin and play of color—whether we take the gems, or the more perishable flowers and fruits of vegetation, or, in fact, any of the objects of natural history—is one of the most delightful yet perplexing inquiries in the secrets of Nature. The variegated hues of the flowers puzzle the philosopher; but they and the prismatic flashes of the diamond are no less interesting and inexplicable than the illumination of that marvel of the insect world, the fire-fly of Jamaica (*Pyrophorus noctilucus*), which displays at the same moment magnificent flashes of green and red light.

The chemist says the rainbow-hues flashing from the transparent gems, and dazzling the eye with their lovely and fugitive play, are only the effect produced by the presence of certain oxides of the metal. "Color," exclaims the philosopher, "is only the absorption and reflection of light, or simply the difference in the rate of vibration of the rays; and that color is not always inherent, but is developed by some extraneous cause." One chemist declares the amethyst to be colored by the oxide of manganese; but another, more particular in his analysis, finds none: yet the glass-workers produce facsimiles by means of that metal. Still another maintains that color is due to the complex structure of certain chemical molecules, and points for instances to the splendid aniline colors, whose royal hues appear one after the other according to the combination of these molecules.

Certain gems change their hues when heated, but return to them when cool, like the spinel and the Saxon topaz. Can the color, then, be considered ponderable? or is it only the molecular arrangement of its crystallization which produces the colored rays? Prof. Stewart asserts that the powers of radiating and absorbing light depend greatly upon molecular condition, and shows that a tourmaline heated to incandescence emits light polarized in a plane perpendicular to that which it transmits. Here the structure that enables the crystal to take up wave-motion in one direction only, compels it to impart motion exclusively in the same direction. Hoffman's celebrated experiments seem to indicate that each particular degree of refraction causes a different set of vibrations, resulting in a different sensation for every part of the spectrum, and reproducing the effect of various colors on the optic nerve. The action of the galvanic battery on the aniline series suggests the idea that the colors evolved are due to nascent oxygen, and that the tint corresponds to the degree of oxidation.

We say that solar light is quite if not absolutely necessary for the production of color, and that the shells of the mollusca, living at great depths in the ocean, where total darkness is supposed to reign, are devoid of any decided hue. Yet there are remarkable examples to the contrary; for most of the zoophytes living at the same depths as the mollusca are nevertheless high-colored animals. The ulocyathus is of a refulgent scarlet, yet lives at the depth of more than one thousand feet; the actinopsis is of a fine yellow; and the capnea displays a decided red tint: yet they flourish and propagate at the enormous depth of nearly two thousand feet below the surface of the ocean. These phenomena may depend upon the food of the animals; and it may come floating down from the upper waters, and produce similar effects to those we witness among animals on land when fed with peculiar food. The beautiful red coral, however, is never found colored below the depth of one hundred and fifty feet; and the finest tinted specimens come from the depth of twenty-one to twenty-four feet.

It was once affirmed that the immense range of color we observe in the almost endless variety of the flowers of vegetation is due to different proportions of the oxides of the metals; and that especially the changing and gorgeous hues of our autumnal foliage were produced by the effects of that chameleon of minerals, manganese. But we now think that it is from the rays of the sun and their effects we derive all the beautiful colors that deck animate and perhaps inanimate nature. They certainly give to the earth the tender blue of the violet, or the delicate crimson of the rose; the gorgeousness of the plumage of the peacock, or the brilliant decorations of the insect world; the lovely green of the meadow, or the purple and golden shadows of the distant mountains. And in those countries where the sunlight is

The Tourmaline

the most constant and powerful, there we find the colorings of animal and vegetable life in the greatest variety and perfection. We may even carry our speculations still farther, and maintain, with a degree of probability, that many of the minerals derive their colors from the same source.

In seeking for the causes which give rise to the multitude of hues observed in the gems, many interesting and mysterious results are obtained in the experiments of the laboratory. These experiments are varied, and almost endless; but at the present time we will content ourselves with reviewing some of the effects produced by heat.

Protoxide of iron, says the chemist, colors the tourmaline, as well as the spinel, chrysoberyl, sapphire, emerald, zircon, garnet, turquoise, chrysolite, and others among the gems. Yet these stones, when exposed to the fire-test, or red-heat, exhibit phenomena which hardly support the general views of the chemist. The vivid as well as the more delicate hues of the red tourmaline vanish like magic almost at the first touch of the blow-pipe flame; while the green and the blue are not much affected at a higher temperature. The charming tints of the emerald are completely destroyed by a red-heat; while the green sapphire and the green diamond are unchanged. The garnet loses its red hues by heat; but the green are preserved. The Saxon yellow topaz changes to white when exposed to a high degree of heat; but the Brazilian topazes, of darker hues, become of a beautiful rose-pink at a low temperature: but, if the heat is continued to a certain degree, neither hue can be recalled, and the stone becomes colorless. The red sapphires often gain a deeper tint; while the bluish in color sometimes become snow-white. Most of the sapphires retain their hues even after having been submitted to a very high degree of temperature. Berzelius found that the red spinel of Ceylon, when exposed to the action of fire, became brown; then black and opaque; but, when cooling, it changed to green; then afterwards became limpid; and finally was restored to its original color. The diamond, when heated, sometimes becomes pink; which color gradually fades away on cooling. The hues of the amethyst and zircon are completely destroyed at a red-heat. The chlorophane from Siberia exhibits an emerald-green when heated: and still another variety from the same locality, when heated to 212° Fahrenheit, phosphoresces, and becomes green; but, when heated to a still higher temperature, it becomes blue. In all cases with the fluorspar, when the phosphorescence ceases, the color of the mineral disappears, and never returns. In the outer and inner flames of the blow-pipe may be seen remarkable effects, which are not clearly explained by the oxidizing and deoxidizing power. Protoxide of iron with the fluxes gives brownish-yellow in the outer light, and a light green in the inner. Oxide of copper with borax gives a green light in the outer flame, and a red in the inner. The oxide of nickel with borax gives a red bead of glass in the outer flame, which becomes white when cool. Binoxide of vanadium likewise gives a yellow bead in the outer flame, which turns to green when cool. The globule of glass formed with borax and titanium is often emerald-green; but, with a greater quantity of borax added to it, the bead becomes red, which passes to blue or white, according to the degree of heat to which it is exposed.

These experiments, which may be continued with varying effect *ad infinitum,* seem to indicate that crystallization or molecular arrangement has something to do with the definition of color; for heat has the property of removing the integral particles which constitute a body to a greater distance from each other. This idea of molecular arrangement giving rise to some of the colored rays is strongly supported by the revelation of the microscope, which exhibits infinitesimal crystals within the transparent gem. A beautiful illustration of the theory is witnessed in the star sapphires, which, when held directly against the light, exhibit white or yellow stars of long slender rays, sometimes in the midst of a deep blue or red ground; but, when held up to the view in any other direction, the star is invisible. This phenomenon is probably due to myriads of minute crystals, which are arranged in a number of definite planes so as to give rise to the stellate appearance. Mr. Lea, in his extensive microscopic researches with the gems, finds many of them composed of multitudes of crystals which are acicular, cuneiform, or plate-like, or triangular in form, with an angle very acute. Some of the gems appear to be composed of bundles and groups of these minute forms; while in others they are widely separated, or apparently absent, or so small as to escape observation. The garnet affords many beautiful illustrations of crystals within crystals; but, as yet, this veteran observer, after a great number of examinations, has not been able to detect any microscopic crystals within the tourmaline.

Artificial light acts strangely with some of the varieties of the tourmaline, and also with many other gems. The beautiful crimson flashes of some of the red tourmalines vanish as if by magic when viewed by candle-light; and the stones become brown, lustreless, and seemingly opaque. The green varieties are heightened in hue, and the blue are unchanged.

The diamond flashes out its dazzling prismatic rays best by artificial light. It gains vastly in intensity and brilliancy; while the emerald loses perceptibly a part of its superb tint and exquisite lustre. The chrysoberyl from Siberia, called alexandrite, is of a dull green by daylight; but by night this color vanishes, and gives place to a reddish amethyst hue. Imagine the surprise and the chagrin of the amateur, who, obtaining an alexandrite for the first time, hastened to exhibit to an old crony the green gem from the Ural Mountains. The candle was lighted, the table arranged; but, when the green stone from Siberia was rolled forth upon the cloth, it had changed into red!

Some sapphires, which are of a lovely blue tint by day, become of a beautiful amethystine tint by night: other sapphires lose their blue color completely by night, and appear black. The greenish turquoise changes to a celestial blue by night.

The explanation of these strange and beautiful transformations has been partially revealed by the spectroscope; and it is now shown that the cause is sometimes due to the difference in the illuminating lights, as is shown in their spectra. The artificial lights produced by gas, oil, or candles, exhibit in their spectra the same range of color which is seen in the solar spectrum; but each color has not the same relative force. As the artificial light from gas and candles has much less blue than solar light, it is evident that the latent red hues of the gem will preponderate over the blue; and hence some of the stones which are blue by daylight will exhibit amethystine hues by artificial light at night.

A beautiful example of this theory is seen in some of the blue sapphires, which display, a superb violet tinge by candle-light.

The Tourmaline

We may, perhaps, explain by the same theory, or by the absence of certain wave-lengths in the illuminating light, the remarkable changes that take place in some of the purple varieties of tourmaline; or we may account for them by that strange property called "fluorescence," which possesses the power of changing the light in which the object is made visible.

The French jeweller Barbot, in speaking of the tourmaline, well says, "that it seems as if Nature had wished to prove to man that she could imitate in a degree almost perfect even that which she had created the most perfect." And, so far as color is concerned, she has succeeded admirably; for she has endowed the tourmaline with most of the colors observed among the gems. From the wondrous hue of the emerald, which may be taken as the type of all the greens, the color of the tourmaline passes down in easy gradations to the dull shade of the plasma; from the fiery and gorgeous red of the Pegu ruby to the opacity of the jasper and porphyry; from the cerulean blue of the Ceylon sapphire down to the intense black of the carbonado. No other gem has such a vast range of color, not even excepting the sapphire or precious corundum; since the suite of corundum greens is very limited, whilst that of the tourmalines embraces all known shades.

It is often stated that the tourmaline has but little commercial value, and is valuable only for scientific purposes; but, among experts, it is regarded as equal in price to the ruby, the emerald, the sapphire, and the topaz, when it resembles those gems. There can be no reasonable objection to this valuation; since the stone is equal to the emerald in hardness, and even superior to it and the topaz in its brilliancy.

This view is also entertained by the eminent Prof. Beudant of France.

The tourmaline, as we have said before, is a compound silicate of alumina mingled with a great variety of other elements, but in slight proportion; and it is from the two elements, silica and alumina, we derive most of our gems. The emerald, garnet, quartz, onyx, sardonyx, idocrase, andalusia, topaz, are all silicates. The sapphire, with its red, yellow, blue, and varied tints, is pure alumina. The opal, with its thousand charming prismatic hues, is only hydrated silica.

This mineral, silica, seems Protean in its forms and combinations. It not only appears in the crystallizations of the earliest periods of the earth's history; but we find it in the petrifactions of recent times, as in the teeth of the rhinoceros, and other extinct animals of the "Mauvaise Terres" of Nebraska, or in the clamshells of the Arkansas. We also find it in the animal kingdom; and the shells of many infusoria are composed of it. We even drink it in the waters of our bubbling springs, and respire it in the floating dust of the air. It is also silently and mysteriously deposited every day in vegetable life; and the same element that adorns human beauty in the emerald, the tourmaline, and many other gems, gives the gloss and the enamel to the bamboo, and strength to the wheat-stalk, to support its ripening grains.

The optical characters of the tourmaline are indeed wonderful, and they belong almost exclusively to this one mineral. Some of its crystals, when viewed perpendicularly to the sides of the prism, appear of a clear and lively color, and perfectly transparent; but, when they are observed in the direction of their axes, the same limpid stones become opaque. Not a ray of light can be made to struggle through them; in some specimens, even when the length of the prism is less than its thickness. There are, however, wide variations to this degree of opacity; and some of the crystals which possess double refraction undergo no change in transparency, no matter in what direction they may be viewed, being perfectly limpid.

In many of the gems, the ray of light falling upon them is refracted, and divided into two distinct rays. This singular property is called double refraction; and it enables us to distinguish many mineral substances, for all do not possess it. We can thus easily detect crystal from glass, the ruby from spinel, and the zircon from garnet, and many others. The tourmaline exhibits this property; but, when cut in thin plates parallel to its axes, it also possesses the strange and extraordinary power of extinguishing, or causing to disappear, one of these rays of light, while the other is preserved.

This peculiarity of absorbing one of the polarized rays of light is taken advantage of by the experimentalist; and it furnishes a useful and valuable aid in the study of optics. The Mt. Mica stones do not polarize light so well as those from Brazil, or the brown varieties from Carinthia. In fact, the light colored from Maine act very feebly when submitted to the test; and it is doubtful if the clear white do at all. The best polariscopes are made from a slice of a dark-green Brazilian tourmaline opposed to another cut from the brown of the Tyrol. Not a ray of light passes through this instrument, unless a body possessing double refraction be interposed between the two plates of tourmaline. If the transparent mineral is rough, and has not been properly polished for examination, recourse may be had to the use of a glass cell containing a fluid of a high refractive power, like the oil of cassia. The stone immersed in this fluid admits the light in all directions, and is then easily viewed through the plates of the instrument.

By means of this polariscope, which is often called the "tourmaline tongs," the expert is able to detect instantly the character of many of the gems, even when polished, and without the labor of estimating other characteristics, as specific gravity, electrical properties, &c. The ruby is thus distinguished from spinel, the zircon from garnet and many other stones, by their difference in refraction. But there are exceptions to the use of this instrument in the estimation of gems; and the experimentalist must be on his guard, lest he pronounce substances to be of double when they really possess but simple refraction. Glass has a tendency to crystalline regularity when heated and cooled suddenly; and consequently acquires the property of polarizing the ray that passes the first plate of tourmaline, and disposes of a part of that which passes in the second. Certain minerals of the cubic system, like the diamond, which generally exhibits single refraction, produce the same result also by reason of a certain forced arrangement; and, on the other hand, some crystals, like the topaz, when cut in certain directions to their optical axes, cease to exhibit the phenomena of double refraction.

It is interesting to examine this wonderful mineral deposit at Mt. Mica, where the tourmaline occurs in such perfect beauty, and to conjecture how Nature constructed these marvellous stones in the very heart of the granite rocks; how she silently built up in the darkness of the miniature caverns, or in the very substance of the granite itself, the transparent atoms of their crystal forms; how she touched them with the fiery red, the lively green, the mellow yellow, the sombre black, or the tender blue; how, at times, she separated these hues in the same crystal as if by magic touch, or blended them together in exquisite transition and gradation. Here, among this grand display of the rare and the beautiful, Steno might have properly spoken of the play of Nature,—Steno, who began geology; whom Deluc called the first geologist.

We may imagine Nature at work creating these gems by the same law by which she constructs the crystals of snow, whose forms are almost endless, but whose every angle is one of sixty degrees. How she does this we may perhaps learn by destroying a block of ice by means of the electric beam, which delicately dissects with infinitesimal touch the structure of the crystal edifice by reversing the order of its architecture. "Silently and symmetrically the crystallizing force had built the atoms up: silently and symmetrically does the electric beam take them down. Here we have a star, and there a star; and, as the action continues, the ice appears to resolve itself into stars."

This process of crystallization is indeed replete with wonders. It is a marvellous play of forces, that enables the molecules of bodies to build up those beautiful crystals whose sides and faces are as polished and symmetrical as though shaped by the dexterous hand of the lapidary. Some rocks exhibit certain forms of crystallization with unvarying restriction; while others are more flexible in modes of construction. Yet there seems to be nothing left to chance. Everywhere fixed laws affect the conformation, and determine the structure, of the whole. The diamond erects itself into a crystal from atoms of carbon; the sapphire is shaped into symmetrical form from atoms of alumina; and the tourmaline is built up of a multitude of substances, all combining in definite proportion.

Although the forms assumed by the different minerals and salts occurring in Nature appear, at first sight, infinitely variable, and independent of all fixed principles, yet the ultimate forms have been reduced to a few. In studying the composition of these gems, we may learn of the remarkable laws which govern all combinations, and startle the inquiring mind with new ideas of the infinite. In many of these complex minerals, man has not yet discovered the law of combination. Yet, with wonderful patience and skill, he has unravelled some of the simpler problems; and he has so far obtained an insight into the workings of Nature, that he knows that certain bodies, simple or compound, combine between them in a determinate and fixed proportion, and admit of no intermediate.

CRYSTAL OF TOURMALINE
Exact Size.

Hamlin Collection. Mt. Mica, ME

Chapter V

Origin of the Precious Stones, etc.

When and how were the gems deposited? and has Nature ceased to reproduce them? is often asked by the inquiring mind. This is a question not so easily answered as it may seem at first sight; and there are many facts known to the mineralogist which incline the observer to believe that Nature (or rather her forces in this respect) is not yet exhausted. There is a widespread belief among the numerous and far-separated diamond-seekers among the gem mines of India, that the diamond is being constantly reproduced; and this idea, though apparently absurd at first glance, is not destitute of plausibility when we come to consider the geological conditions of the gem districts, and the phenomena of terrestrial magnetism which are observed there. The artificial production of a great variety of minerals among the slags of furnaces affords corroborative evidence; and we may see to-day the opal and hyalite forming from the decomposed cement of the Roman ruins in the hot-springs of Polombières. Nature, evidently, still possesses the power. Certain conditions alone are wanting.

The mysteries of this creative power have not yet been solved by science; but investigations thus far go to show that the ancient astronomers—Hermes, Ptolemy, and others—were quite right when they maintained that all things here below were governed by the influence of things above. It is perhaps true, that the grand geological changes of the earth have ceased; but, nevertheless, the mighty and mysterious forces that deposited the metals, the gems, and arranged the formation of the conglomerate rocks, or those of a high molecular organization, are yet at play. Every glancing sunbeam contains a world of power within itself; and its effects are visible and felt on every side. In the tiny columns of dust whirling in the road, the tall sand-pillars sweeping majestically over the desert, the water-spouts of the ocean, the terrible cyclones of the topics, and in the devastating earthquakes,— "those thunder-storms in the earth,"—we witness phenomena due to the same Protean and wondrous force,—electric action. "The brilliant folds of the aurora—that magic arch, surpassingly beautiful by the brilliancy of its colors, mysterious from its swift and silent play across the heavens, flaming as with the glow of unseen and unearthly fires"—are also due to the same influence, or terrestrial magnetism.

All these wonderful changes and results are not produced by the earth itself, but are due to extra terrestrial influences,—the cosmical action of the surrounding bodies in space, the varying action of the sun, moon, and stars. From the sun proceed these strange impulses; and, in fact, the whole electric system of our globe quivers, as it were, continuously under the influence of the solar forces. All motion on the earth, all life, even death itself, comes to us in the sunbeams. We must not, however, confound electric force with the creative power. Electricity is not life: it is only the instrument of life, and nothing more. This subtle agent is ever present on earth, and never ceasing in its action. Not only do we witness its marvellous play in inanimate nature; but we may find it constantly at work in our own bodies. It performs a part in the perfection of human thought and physical beauty, as well as in the vitality of the vegetable world or the delicate forms of crystallization. We little know of the vast energies of this unseen power, which are silently and constantly exerted around us in every chemical change.

Where the skies are the fairest, and the sunshine the most constant, there the electric disturbances are the most powerful, and there we find the magnificent colorings of animal and vegetable life in the greatest perfection and variety. Even the clouds that clamber up the sides of the lofty peaks of the Andes, are observed to assume richer hues and shades when they approach the trachyte ledges. For a long time, when most of the gems brought to the commercial markets came from Southern Asia or the equatorial regions of America, it was thought that the heat and light of the tropics were required for their production; but we now find them, though in limited quantities, among the snows of Siberia, and beside the glaciers of the Alps. However, it is a singular fact, that many of these gems in a remarkable degree lack the tints and lustre of those found under the sun of the tropics. There is, especially, a great difference observed in the spinels of Siberia and North America, and those of Ceylon; the former being dark and impure in color, while the latter possess all the most brilliant characteristics of the species. Naturalists observe the same marked differences in color between the fishes, the birds, insects, and animals of the cold and warm zones. These phenomena are thought to be due to solar influences; and, upon like hypotheses, the magnetic currents ceaselessly playing through the crust of the earth, may, perhaps, to-day produce the diamond in certain favored localities, thus realizing the belief of the Hindostanee when he discovers new gems in the oft-washed sands. In these favored localities, certain elliptics, where the electric forces are in active play, may be readily traced in marked distinction to other territories; and it is impossible, in the observance of their varied effects, not to connect with them the production of the gems in past times, if not at the present day. How Nature accomplishes these results is yet involved in mystery; but we may glean new ideas from the artificial crystallizations that take place in our laboratories under the action of electricity. Furthermore, may not this Protean agent exalt the properties of atmospheric oxygen, and act like a germ in its transformation of hydrogen into an agent of decomposition and combination? In studying the mysteries of metals, Graham was tempted to believe, from the action of hydrogen in its occlusion by certain metals, that the gas is itself a metal, possessing affinities for other metals; and so firmly was this view fixed in the chemist's belief, that he gave to hydrogen the name *hydrogenium*, considered as a metal. The formation of fulgurites, which shows that the lightning may melt and fuse without evolution of heat, is also to be considered in connection with the study of this subject.

Crystal of Tourmaline
Mt. Mica, ME

The History of Mount Mica of Maine, U. S. A.

and Its Wonderful Deposits
of Matchless Tourmalines

Augustus Choate Hamlin, M.D.

(1895)

"It is a strange analogy, well worthy of fixing the attention of philosophers. These jewels, which have the privilege of attracting our gaze, and of fixing our eyes upon them by an unaccountable species of magnetism, appear also to incite the secret affinities of lightning."

—Abbé De Tonville.

Chapter I

Mount Mica is situated in Paris, the shire town of Oxford county, in the state of Maine. The village is one of those quiet, secluded places, enriched by political and scientific tradition, and adorned with scenes of great natural beauty, which the poet and the philosopher seek for, and cherish when found.

A mile or more to the eastward of its court house, at the foot of the slope of a mountainous ridge which like a great mural wall hides the eastern horizon from view, appears a little hill seeming to be one of the buttresses of the rocky range towering above it. This little hill, with its great gray piles of broken rocks, which have been torn from its surface by the explorations of more than half a century, disfiguring its natural beauty, is the famous Mount Mica, known to every mineralogist of note in the scientific world, and dear to every heart that loves and respects the beautiful and the mysterious among nature's works.

It was discovered in 1820 by two students who had become interested in the study of mineralogy, and who spent much of their leisure time in searching for minerals among the exposed ledges and the mountains around the village. Late in the autumn of 1820, and on one of its clear, calm days, they started out to explore the range of hills which form the eastern boundary of the town, and stretch away to the northwest, until lost among the mountains around Molly Ocket. The names of these two students were Elijah L. Hamlin and Ezekiel Holmes. Hamlin was a resident of the village, but Holmes was a visitor, and temporarily a student, in the place. They had spent most of the day along the mountain ridge to the southward, and were descending the western declivity on their way home, just as the sun was setting behind the great White Mountain range, fifty miles or more away on the western horizon. At this moment the view of the intervening country, diversified in color and in shade, together with the gorgeous masses of changing clouds in the western sky, formed a picture of great beauty, and young Hamlin, fascinated with the entrancing picture spread before him, halted for a moment on the crest of a little knoll to enjoy the scene. On turning to the eastward for an instant for a final look at the woods and mountains in his rear, a vivid gleam of green flashed from an object on the roots of a tree upturned by the wind, and caught his eye.

Advancing to the spot, some rods in his rear, he perceived a fragment of a transparent green crystal lying loose upon the earth which still clung to the root of the fallen tree. The student clutched the glistening gem with eagerness, and called back his companion, who had already passed over the brow of the hill, and was some distance down the slope. After examining the newly found treasure, the students carefully searched the surrounding soil for other specimens; but the rapidly increasing twilight soon compelled the youthful mineralogists to abandon the search. They, however, resolved to return early in the morning and continue the exploration. But during the night a storm arose, and covered the hill and its

Panoramic View of the Excavations at Mount Mica

adjacent fields with a thick mantle of snow, which remained until the next spring.

As soon as the winter's snows had melted away, and had left the hill and its sides exposed to view, the students again returned to the search, and this time with success. They went directly to the bare ledge which crops out on the brow of the hill, and which they had not examined on their previous visit; before darkness had overtaken them. As they climbed up over the smooth and denuded surface of the rock, they were astonished to observe many crystals, and fragments of crystals, lying exposed upon the bare ledge, and sparkling in the rays of the sun. These they carefully gathered; and tracing others to the earth below the ledge, and which had formed from the decomposition of the rock, they eagerly turned up the soil in search of its hidden treasures. Thirty or more crystals of remarkable beauty and transparency rewarded the labors of the students, and with mingled feelings of joy and wonder they held them up to the rays of the sunlight, and admired their varied colors of green, red, white and yellow in different shades. They had, indeed, stumbled upon one of the richest and rarest of nature's laboratories. All around the brow of the ledge, enormous masses of rose red lepidolite, splendid groups of crystalized quartz of white or of smoky hues, black crystals of the oxide of tin, broad foliæ of glistening mica, snowy flakes of feldspar, studded with minute transparent crystals of green and red tourmalines, lay scattered about in profusion. Collecting as many of the choice and beautiful specimens as they could carry, the students, heavily laden, returned to the village, and sought to ascertain the nature of their mineral treasures. Subsequent examination indicated that the ledge was perforated with cavities, in which the tourmalines and other minerals had been deposited. It was also evident that the crystals that had been gathered by the students had been set free from their cavities by the decomposition of indefinite periods of time, which had disintegrated the surface of the ledge. There was no evidence of drift; and the crystals lay exposed upon the rock, while the softer and more perishable materials had been set free and washed by the rain, and blown by the winds, down to the base of the ledge, and accumulated in time as soil. Parts of the ledge yet exposed to view were fairly honeycombed with small cavities and soft spots, where the decomposing feldspar was crumbling away. In these cavities and decayed places in the rock, other tourmalines were obtained by breaking away the edges of the cavities, or removing the decomposed material.

The discovery having been made known to the villagers, many of them hastened to the place, and secured a number of fine specimens as trophies or as souvenirs. As no one in the vicinity was able to distinguish the character or nature of the specimens, or even to call them by name, the students enclosed a few of the smaller crystals in a letter to Professor Silliman, of Yale College, and requested him to describe them. He kindly and promptly informed the youths that the minerals were tourmalines, and of rare occurrence. Thereupon the students selected some of the finest and purest of the crystals, and addressed them to the professor, in return for his kindness. The parcel was entrusted for safe keeping to the late Governor Lincoln, who was then a member of Congress, and about to start for Washington. At this period the journey to the capital was a serious undertaking, and the condition of the roads was such that most of the distance had to be traversed on horseback. The Governor started safely with the precious package, but lost it before reaching New Haven, and no trace of it has ever been found.

Two years after the discovery, the two younger brothers of the discoverer, Cyrus and Hannibal Hamlin, although scarcely in their teens, resolved to make a more complete exploration of the ledge. Having borrowed some blasting tools in the village, they proceeded to the hill and managed in a rough way to drill several holes in the ledge and blast them out. These operations, though of trivial magnitude, were attended with unlooked-for results, for the explosions threw out, to the astonishment of the boys, large quantities of bright-colored lepidolite, broad sheets of mica, and masses of quartz crystals of a variety of hues. The last blast exposed a decayed place in the ledge, which yielded readily to the thrusts of a sharpened stick or the point of the iron drills. As the surface was removed, great numbers of minute tourmalines were discovered in the decomposed feldspar and lepidolite. The rock became softer and softer as the boys proceeded in their work of excavation, and soon they reached a large cavity of two or more bushels capacity. This hollow place, or rotten place, appeared to be filled with a substance resembling sand, loosely packed. Amongst this sand or disintegrated rock, crystals of tourmaline of extraordinary size and beauty were found scattered here and there in the soft matrix. Scratching away with renewed energy, the boys soon emptied the pocket of its contents, and found that they had obtained more than twenty splendid crystals of various forms and hues. One of these was a magnificent tourmaline of a rich green color and a remarkable transparency. It was more than two inches and a half in length by nearly two inches in diameter, and both of its terminations were finely formed and perfect.

Several others possessed extraordinary beauty, and some of them were quite three inches in length and an inch in diameter. The colors of these tourmalines were quite varied, but were chiefly red and green, and equaled if they did not surpass in beauty the crystals previously collected by their brother Elijah. The exact number of crystals obtained by the boys is not known; but when collected together, with the fragments of others, they filled a basket of nearly two quarts capacity. Besides the tourmalines, the quantity of lepidolite, mica, and other choice minerals thrown out by the blasts or found in the sides of the cavity was so great that the boys were obliged to seek for an ox-team to transport them home. So little was known of the value of these rare minerals at that time that the possessors considered the finest of their treasures to be worth about a guinea. Cyrus afterwards learned from his brother Elijah, who had moved to the eastern part of the state, the names of mineralogists in the United States, and also in Europe, who had made inquiries of him concerning the discovery of Mount Mica, and the disposition of its minerals. With some of these he placed himself in correspondence, and from time to time disposed of nearly all of the finest of the crystals in exchange for money or for minerals.

Cyrus not long afterwards moved to Texas, where he died many years ago, and with him has perished the history and the distribution of these remarkable crystals and gems. The younger brother and survivor, the late Vice President Hannibal Hamlin, took but little interest in mineralogy, and gave his share to his brother Cyrus. He remembered, at the time of the publication of the treatise on the tourmaline, in 1873, only the facts of the discovery, the curious and symmetrical forms, the perfect limpidity and the wonderful beauty of the crystals.

This is all that is known of the history of the splendid gems and the remarkable crystals that Mount Mica yielded to the explorer in its early and best days. Gathered then in profusion, and carelessly treasured, they have since been scattered over the world, and in many instances their identity has been completely lost.

A very fine and perfect crystal of light green tourmaline, three and a half inches in length, was loaned by Elijah in 1822 to a friend in Portland, and was lost in that city. Another beautiful specimen was given to Governor Lincoln, who afterwards gave it to the lyceum at Worcester, in Massachusetts, where it remained for many years, until it passed into the hands of Professor Burbank, by whom it was cut. The Lapidary Reynolds obtained from it a splendid gem of great beauty and purity, of fifteen karats in weight. A few days afterwards Burbank lost it while crossing a field of grass land, and all efforts to recover it proved unavailing.

The late Professor Cleaveland, of Bowdoin College, a famous mineralogist in his day, received several choice crystals, and among them a superb yellow tourmaline, which was given him by Elijah Hamlin. But there is now no trace left of these specimens. His cabinet, which Bowdoin College inherited, does not now contain them, but from evidence gleaned from his correspondence, it is surmised that they may have been sent to his friend, the celebrated

Berzelius, and it is possible that some of them may now be in the mineralogical cabinets of Sweden. Those which fell to the share of young Holmes at the time of the discovery were destroyed many years ago in the fire which burnt the Gardiner Lyceum, and there is no description of them to be found. In the imperial collection of minerals at Vienna, there were said to be some tourmalines of remarkable beauty, and mineralogists were struck with their resemblance to the Maine tourmalines, especially in their arrangement of color. They came from the cabinet of the famous antiquary Van der Null, and were simply labeled "America." This is all that is definitely known of their history.

As the tourmalines of all known localities have peculiarities which distinguish them in a marked degree from each other when viewed by the practiced eye, it is sometimes easy for the mineralogist to give the locality to the unlettered specimen. And from the evidence submitted to us, we feel quite confident that some of these tourmalines at Vienna are a part of the results of the early explorations of Mount Mica. Baron Lœderer, an experienced mineralogist, happened to be in Vienna when the Austrian government purchased the collection of Van der Null. He was present when the boxes were opened, and as he had visited Mount Mica previously, and was acquainted and familiar with the characteristics of the mineral, he at once recognized them to be identical with the tourmalines of Maine, and believed them to have been taken from Mount Mica previous to the year 1825. The baron, before his death and previous to 1830, gave his views on the subject to Dr. Louis Feuchtwanger, the writer on gems, and by him this information was communicated to the author.

In 1891 the Vienna cabinet was examined by the able expert on precious stones, George F. Kunz, of New York, and he reported shortly after to the writer that the collection of tourmalines in this imperial cabinet was far below some of those in America, and were not to be ranked as among the best specimens of Mount Mica. Mr. Kunz has been familiar with the finest crystals found at Mount Mica, and preserved in the cabinets of the United States, and as he has examined most of the great collections of Europe, his opinion is not to be questioned. The tourmalines in the cabinet at Vienna may have been changed greatly by neglect or exposure during the last fifty years, as has been noticed with some of the fine specimens in this country.

On November 20, 1822, E. L. Hamlin sent to Professor Silliman, at Yale College, a box of minerals from the locality, with letter and catalogue as follows:

> Dear Sir: I herewith transmit to you, by the hands of Hon. Enoch Lincoln, a small package of minerals, and am in hopes that I may soon have an opportunity to send on a box of more numerous and larger specimens. Paris, the shire town of Oxford county, Maine, has been settled only about forty years, and the country around it is comparatively yet a wilderness; and until within about a year there has never been any examination made of its minerals, and the only search that has been made within this time has most richly rewarded the labor. Most of these minerals, as you will perceive, have the same locality, are

found near by, and were discovered a short time since by Mr. Ezekiel Holmes, a student in medicine at this place, and myself, while on a mineralogical excursion.

This place seems to resemble much the Haddam and Chesterfield localities, inasmuch as it contains a similar colored mica, and embraces nearly, if not entirely, the whole family of the tourmaline. The country around here, elsewhere, seems to be peculiarly rich in minerals.

In the catalogue and description which accompanies the letter, Mr. Hamlin states that specimens of green tourmaline have been found there from one-eighth to one and one-half inches in diameter to one and even six inches in length, perfectly transparent, and of the lightest to the deepest green. He also describes radiated and acicular green tourmalines in small prisms, and from one to six inches in length. Also indicolites, passing from blue into green and red, black or brown. Also white tourmalines, the most of which are tinged with red. The red tourmalines he describes as mostly found in the green crystals, or encrusted with the green on its surface. He describes them as very beautiful, and varying from a pink to a deep crimson. This letter and catalogue was overlooked by Prof. Silliman for some time, and was not published until February, 1826, in *Silliman's Journal.* In the catalogue Mr. Hamlin mentions lilac colored mica as found in small globular concretions, consisting of minute foliæ, placed one upon another so as to form short columns, situate mostly parallel to each other and held together by a siliceous cement, etc.; this is evidently what was recognized later as lepidolite. Furthermore, he states that "all these minerals are found at the same place, on top and on the declivity of a small hill, its surface measuring perhaps one acre, and elevated not more than forty or fifty feet above the land around it. The base of the whole hill is probably a ledge, but it breaks the surface only on the top in a space of about four rods square, exhibiting a ledge of coarse granite, thickly filled with mica and tourmalines, of which the black principally predominates. But little search has been made, and only in one place have we gone under the surface; and it was there that we found the best specimens loose in the soil."

In 1823, Prof. Webster, of Boston, learned of the deposit, and among the letters of Elijah Hamlin we find the following, written to him on the 28th of April, 1823:

> Dear Sir: I have for some time been hoping to be able to visit Portland and Brunswick, and on my way there should be exceedingly happy to examine the spot where you obtained the specimens I had the pleasure to receive from you through Dr. Eaton. I know not whether the box I sent you arrived safely, as I have not had the happiness to hear from you since I sent it. In the first number of the *Boston Journal of Philosophy and the Arts,* which will appear early in May, I have taken the liberty of noticing these specimens, and announcing you as the discoverer of

them—am I right? Will you inform me if there is a prospect of obtaining any better specimens should I visit Paris? I am ready and shall be most happy to send you any specimens you may desire in exchange for those of your vicinity. I am, very truly yours,
J. W. Webster.

We have further evidence that the ill-fated Prof. Webster came twice to Paris and explored the ledge to some extent, but the time of his visits is not known. The miner who worked for him told the writer, more than twenty years ago, that at one time he obtained a fine green crystal of considerable length, and at the last exploration he found a red crystal of great beauty and purity of color, and as large as the thumb.

Webster became greatly excited, as the miner handed him the crystal from the bottom of the pocket, and he ran to the top of the ledge, holding up the gem in the rays of the sunlight, and, dancing around like a madman, he exclaimed to the amused miner that he would not take a large sum of money for it. No farther trace can be found of these and other specimens that Webster was known to have, and it is surmised that they were probably sold to or exchanged with European mineralogists.

July 1, 1824, L. Willis wrote Mr. Hamlin that he could not find a lapidary in Boston or in Salem to cut the stone sent to him, but there was one in Providence who could do the work. He further adds that Prof. Nuttal, of Cambridge, would visit Paris in the course of a few weeks for the purpose of viewing Mount Mica, etc., but beyond this letter we can find nothing relating to the results of this visit.

Under the date of August 10, 1824, we find a letter from John Pedrick, of Salem, stating that he had seen the fine gem which Willis had in his possession, and desired Mr. Hamlin to send him one to be set as a bosom pin for himself and his children. He also mentions the fine specimens of green tourmaline which he had received from Mr. Hamlin, and further states that Prof. Nuttal had started for Franconia and Paris.

Chapter II

In September, 1825, the distinguished geologist, Prof. Charles Upham Shephard, of Amherst College, was attracted to the place by the reports made by Mr. E. L. Hamlin and Dr. Holmes, and in the July number of *Silliman's Journal*, 1825, he published an account of his visit and the results of his explorations, which are too interesting and too important to be overlooked in this memoir. He states in the *Journal* that "the locality in question is situated upon the farm of Mr. Chesley, who lives upon the road leading from Paris to Buckfield, one mile east of the village of Paris. The rock of the vicinity is graphic granite, and, except where it breaks through the soil in large ledges, as it does in several places, it is very much shattered by decomposition. This is particularly the case upon the farm of Mr. Chesley. In the field where the minerals occur, angular fragments of graphic granite of all sizes are seen protruding above the surface of the ground; and in digging, we find apparently a soil just formed, consisting of gravel mostly derived from feldspar, and as yet but slightly discolored with vegetable mould. In the highest part of the field, and just in front of a little wood, the granite makes its appearance in a continuous mass for the compass of a few square rods, and it is here possessed of a high degree of integrity. Such,

View of Mount Mica from the Maxim Hill, on the North

however, is the abundance of foreign substances which it contains, that its graphic character is no longer obvious. It is here that the tourmalines and other minerals occur. When the locality was first visited, large masses of lepidolite, in some instances entirely coated with rubellites, and loose crystals and fragments of crystals of the differently colored tourmalines, together with groups of crystallized quartz, were found dispersed over the surface of the hill. These, however, have long since disappeared; and the collector who is now in search of these minerals is obliged to lay open the solid rock by the aid of gunpowder. The granite is composed chiefly of feldspar, and on this account is the more easily quarried. It is traversed by several irregular veins of mica and lepidolite, the latter of which are occasionally nearly a foot in width.

"These veins, as well on account of the mica and lepidolite as the substances they embrace, are the principal objects of pursuit with the mineralogist. The mica forms veins of six or eight inches in width, and exists in large foli, among which small portions of quartz and feldspar are interfused. When detached, it presents imperfectly formed rhomboidal crystals, with a tendency to the figure of the mica prismatique. Some of these attain a foot in length and seven or eight inches in breadth. In general, however, they do not occur in pieces above half of this superficial size, and with a thickness of about one inch, the lamina composing them being straight and closely aggregated. When held between the eye and the light, with the prismatic axis toward the eye, the light that is transmitted is faint, and of a rich, reddish brown color; but on giving the crystal a revolution through half a circle, more and more light is transmitted. It is in a position nearly perpendicular to the axis when the light penetrates the crystal most freely, and this notwithstanding the quantity of matter through which it is obliged to pass in the latter position having become considerably augmented, the light continually changing in color as well as in intensity, and finally becoming of a greenish yellow tinge. This mica, although interesting on its own account, is still more so on account of the tourmalines which it embraces, and which are disposed in long, acicular crystals between its laminæ. The largest of these are about a quarter of an inch in thickness and three or four inches in length. They are for the most part of a leek green color, and transparent. They are rarely isolated, but much more generally variously grouped, etc., etc. The lepidolite of this place seemed to us also very interesting, from the abundance in which it occurs, the variety of tints it offers, and the beauty of its imbedded minerals. Little difficulty, I imagine, would be experienced in obtaining pieces one foot in diameter. Its colors go through every possible variety of peach blossom red, from the deepest tint to that which is the palest. Its composition is granular, consisting of imperfect hexagonal concretions of various sizes, from that of a pepper corn to a pin's head, which are intimately and confusedly aggregated, often with an intermixture, in the deepest-colored specimens, of transparent quartz, the lepidolite completely penetrating the quartz. Masses of the last description are broken with the greatest difficulty, being surpassed in toughness by no mineral with which I am acquainted, excepting perhaps nephrite and petalite. This variety, I am persuaded, would prove exceedingly beautiful if cut and polished, and must resemble the finest avanturine *a pluie d'argent*. Like the lepidolite of Rozena, in Moravia, it contains crystals of rubellite, which, though less abundant, are perhaps more remarkable for their size and delicacy of color. The paler varieties of lepidolite, which are more free from quartz, but which contain occasional admixtures of Cleavelandite feldspar, afford the most delicate crystals of this mineral. They are tolerably perfect six or nine-sided prisms of about one inch in length, and possessed of a very delicate rose color. The deeper colored lepidolite, on the other hand, in those parts where the quartz and feldspar predominate, afford occasionally large crystalline masses of the same colored rubellite one or two inches in diameter, and sometimes in lengthened prisms, inclosing indicolite of an intense blue color and a somewhat conchoidal fracture. The sea green colored tourmaline accompanies more rarely the above mentioned variety, but none of them occur in pieces sufficiently exempt from flaws or endued with the requisite transparency to entitle them to the character of gems, like the specimens described in the sequel."

He also mentions the substance which he found in the quartose cavities among the lepidolite, and which often coated the crystals of the tourmalines. This he then called crystallized white talc, but which is known at the present day as Cookeite. He also alludes to the beryls and the minute crystals of zircon, and then proceeds to describe his researches. "I commenced my researches directly upon the top of the tourmaline ledge, not, indeed, in the firm granite, but in a covering of loose materials reposing upon it to the depth of four or five feet. Here a slight digging had been commenced, over a surface of a few feet, apparently in search of the fine crystallizations of brown quartz with which it would seem that this particular spot formerly abounded. On causing the exploration to be renewed, an abundance of this substance was thrown out; and very soon I began to meet with masses of lepidolite, completely studded over and penetrated by finely colored crystals of green and red tourmaline and drusy fragments of granite, whose cavities were lined with the same material minerals—the feldspar being nearly opaque, of a delicate whiteness, and possessing the beautiful chatoyment which this species often presents; crystals of greyish white quartz, several inches in length and thickness, and penetrated by tourmalines; and, finally, loose crystals of tourmaline and rubellite from a quarter of an inch to two inches in diameter. Thus we followed digging in every direction so long as it continued to afford these products, which it did until within a short distance of the rock. The majority of pieces, however, seemed to occupy a vein one foot wide and three feet long by about two feet in depth. From this state of things, it seems fair to conclude that the granite here, when in a state of integrity, must have possessed a drusy cavity open from above, and it is by no means improbable that the loose specimens of tourmaline, smoky quartz, etc., which were found about the sides of the hill on the discovery of the locality, had their original repository in this cavity." In this article he alludes to the forms of crystallized quartz, both clear and smoky, which were found in the cavity, and from his observations he was led to conclude that the brown quartz was first deposited from solution, and surrounded the walls of the granite with its crystals, and that the tourmalines crystallized next, to which succeeded the talc and feldspar; and that, finally, the white quartz was deposited around the other substances. In this cavity which he uncovered he found many remarkable crystals of tourmaline of various colors, some of which he gives a description of, but it is evident that he did not attempt to restore the broken crystals with the fragments, which were undoubtedly found at the same time, and thereby estimate the full beauty and form of the crystal as it appeared in its perfect state, and has been done with great care by the explorers of the past twenty years. Among those found at this time he mentions several, as follows: No. 1 he describes as being 1½ inches long, with a diameter equal to its length, and having one termination with, polished faces. Its color was an intense grass green, with a tinge of blue. No. 2 he states as the fragment of a crystal 3/4 of an inch in length by 1.2 of an inch in diameter; the extremities he polished flat. Its colors were green, passing into pink. No. 3 is a crystal 2 inches long by 1½ inches in thickness. Its sides are coated with green tourmalines to the depth of about a line—the whole interior, from end to end, consisting of the most beautiful rubellite. The color is more intense at one extremity, and is deepest throughout at the center. One end is of a dark and exceedingly rich blood-red color, becoming slightly amethystine towards the circumference, while the other approaches more the color of a crimson in which little if any blue is discernable. No. 4 is a crystal 2½ inches by 1 inch. Viewed across its axis, at one extremity it exhibits a fine sea green, while at the other it is of a rich crimson red. A joint detached from the green end presents, when viewed in a line parallel with the axis, a grass green, bordering on a pistachio green color. No. 5 is a crystal measuring 1½ inches each way. Its color, when viewed across the prism, is a sea green, with a large proportion of blue; but it passes into a pale rose at one extremity. A brooch was cut from the green end of this crystal which measures 19-20ths of an inch long, 16-20ths broad and 8-20ths in thickness. It is cut after the manner of a large emerald. The large plane forming the front face, and which is situated at right angles to the prismatic axis, is 2/3 of an inch in length, by a little more than half an inch in breadth. Its color is intermediate between grass green and pistachio green, and its transparency is perfect. It contains but one flaw, which is invisible when the brooch is held in ordinary positions. No. 6 is a section of a prism about 1 inch in length by 2 inches in diameter, of a pale pink color, except a thin coating, which is green. This crystal is an exception to the others here enumerated as regards its transparency, freedom from flaws, and beauty of colors, and is noticed only on account of its magnitude. From one of the fragments he had cut a remarkable gem 3/4 of an inch in length, 13-20ths of an inch in breadth, and 4-10ths of an inch in thickness. When viewed by transmitted light it reminds one of one of the finest Syrian garnets; but seen by reflected light, it gives much of the crimson red peculiar to the oriental ruby. Its freedom from cracks, united to its transparency, luster and beauty of color, have caused it to be much admired. He also mentions several prisms precisely resembling the tourmalines from Brazil, and also some indicolites of deep blue color, from one of which he had cut two tables 5/8 of an inch by 1/2 inch. The color was fine, but the beauty was marred by several flaws.

This is all that has been preserved in history concerning the remarkable find of Prof. Shepard, and which was one of the richest yet disclosed at Mount Mica. It is evident that no attempt was made to unite the broken fragments and try to form the crystal again and restore its primitive form and beauty, and it is also evident that all of the specimens were not described, as he does not mention the choice section of green tourmaline which was unearthed at this time, and given to Miss Eliza Hamlin as a souvenir of his visit and his successful exploration at Mount Mica. This specimen, of fine green hue, about 1¾ inches in length by 1¼ in width, is now in the cabinet of Samuel R. Carter, at Paris. The fame of Mount Mica spread among scientific men, and many people came to visit the locality. The Russian and Austrian consuls at New York, Baron Lœderer and Mr. Cramer, were among the visitors, and they carried away with them a large quantity of minerals. Prof. Chas. T. Jackson, the state geologist, when making his surveys, examined the locality, but did not dig or blast so as to test the value of the deposit. But while there he obtained from Mrs. Bowker a fine specimen of light green tourmaline, flawless and of perfect limpidity, which she had picked up in the soil a few days before near the top of the ledge described in plate No. 5.

From time to time during a period of thirty years or more, many mineralogists and collectors of minerals visited the locality and examined it in a superficial manner. In 1863 Prof. Sanborn Tenny, of Williams College, visited Mount Mica, and while prying about the ledge with a crowbar, he accidentally found a decayed spot, which led to a cavity in which he found some choice tourmalines. Among them was a large rubellite or red tourmaline 14 ounces in weight, and probably at that time the largest specimen of the kind known. It was not well crystallized, and was broken in two parts. One of the ends had a thin coating of green, but the rest of the mass was of a translucent dull crimson hue. The pocket also yielded a beautiful section of a crystal 1¾ inches in length by 1¼ in diameter. Both of the terminations had become disintegrated, so that the original length of the prism was not determined. The base of the crystal was of a light crimson, while the upper two-thirds was of the finest grass green and of the purest transparency. (*Vide* plate No. 5.)

Chapter III

For a number of years but little was done in the way of exploring the ledge. In 1864, Mr. Samuel R. Carter, of Paris, commenced work in front of the pit made by former explorers, and started a cut in the ledge 40 or 50 feet to the westward, intending to strike the mineral belt at a greater depth than had been reached, but after removing many tons of rock and finding no signs of the deposit, he stopped work. Some curiosity-seekers, working with Mr. Bowker, then proprietor of the farm, struck a large cavity, and obtained many beautiful fragments of tourmaline of various tints and colors, but they were carried away, and all authentic information about them has been lost.

Shortly after the close of the war the writer, with his father, visited the ledge with the view of renewing explorations if the signs gave any hope of success. At this time the pit was only a few feet in extent and about five or six feet in depth. There was no sign of any of the lithia minerals in sight, with the exception of a minute speck of lepidolite in the southern wall of the pit. The crowd of mineral seekers had carried away all that could be found by digging in the soil and removing all the broken parts of the ledge. Without much hope the writer attacked the ledge at the spot where the speck of lepidolite appeared, and at the first blast was rewarded with the sight of a small cavity which contained a little tourmaline, green in color with a touch of crimson at the base. (*Vide* plate No. 6.) This little cavity was the means of the explorations which have brought to light some of the most remarkable treasures of the locality, and made the place famous in mineralogy.

In 1871 the author determined to test the ledge again, with but little hope of success, however. Prof. Joseph Leidy, of Philadelphia, who had become greatly interested in the deposit, was invited to be present and assist in the operations. Blasting was commenced in September, and after a few tons of rock were removed a large cavity was uncovered, and the large achroite crystal which is now at Harvard was soon unearthed. Prof. Leidy was then invited to try his luck in the cavity. After removing some of the loose Cookeite and broken fragments of the surrounding rock, he was gratified in bringing to light another crystal of white tourmaline which was a little smaller than the first crystal and not so long; nevertheless, a remarkable specimen, and one of the largest achroites known. The first crystal taken out is a complete crystal, white at the top, changing into a smoky hue towards the base, and which assumes a crimson tint when viewed along the line of its axis. The crystal is 4½ inches in length and 1½ in diameter. Both terminations are tipped with green tourmaline. (*Vide* plate No. 8.) A number of little crystals of white tourmalines were also found in this cavity, but none of much value. Nature seemed to have expended her force in making these two fine crystals, and quite all of the specimens were white. Nearly all of the cavities have distinctive colors peculiar to each cavity, and sometimes in a marked degree.

Augustus Choate Hamlin

Later in the year another attempt was made to test the deposit, and after blasting the wall in the rear of the pit to a depth of six feet, the writer was rewarded by finding a little cavity, not much larger than the two hands, but it contained the beautiful crystal which is shown in plate No. 7. The cavity contained nothing more, and its walls were also destitute of any minerals of the lithia group. It seemed as if nature had thrown all her energies in this neighborhood to construct this beautiful crystal, and there was nothing left to form other minerals. In other parts of the ledge, we have noticed cavities to be entirely destitute of minerals of interest, while the surrounding walls and adjacent rock were filled with lepidolite, amblygonite, spodumene, etc., appearing as though nature had exhausted her force in forming the minerals in the outer walls, and had nothing left in material to form in the cavities the beautiful crystals for which the cavities seemed to have been created. Other cavities, during this year, were opened at the depth of six feet from the top rock, but their contents were found to be decomposed by the action of the elements. The water, the frost, and other forces had at this depth exerted their mighty force upon the beautiful crystals of tourmaline, and had rent their solid and transparent forms into numberless fragments. The crystals lay in their sandy beds, undisturbed in regularity of outline, but they crumbled away as soon as touched. Here a summit of a crystal with faceted planes would be preserved, and there the base or a nodule from the central portion would alone remain among the wreck of the marvel of nature's work. The base and sides of these cavities were composed of feldspar and quartz, mixed with lepidolite and other firm minerals, forming natural basins, into which the water trickled down from the ledge above, through its numerous crevices, and so the tourmalines were constantly exposed to the action of water, frequently freezing and thawing, until the walls of the cavities became rent, and the water was allowed to escape to deeper outlets. The appearance of the ledge was so devoid of the lithia minerals that it was thought best to suspend operations for a season, and a party of explorers searching for mica for commercial purposes then took possession of the ledge, and proceeded to remove the rock on the eastern side of the pit. They removed about three hundred tons of rock, and descended to the depth of eight feet, when they struck five well-defined cavities, on a line ranging from east to west, but not connected with each other. The ignorant miners pocketed some of the brighter fragments of the broken tourmalines they saw in the cavities, and shoveled many others away in the dump. Some time after this occurrence, the attention of the writer was called to the discovery, and he made earnest efforts to collect the fragments, and restore the crystals to a semblance of their former beauty, but not with much success. However, the fine crystal which is represented in plate No. 9 is one of the results of his efforts. Among the debris were found the fragments of one of the most remarkable groups of tourmalines yet discovered in any part of the world. They were deposited upon a mass of white quartz about eight inches square and five in depth, and upon this matrix nature deposited nine distinct transparent tourmalines of great beauty and value. Upon its summit arose a crystal of tourmaline quite two inches in diameter and over two inches in height. It was transparent, pink at its base, changing towards the summit to a delicate and gorgeous carmine of a totally different tint from any the writer has yet seen from this locality. On one of the sides of the mass of quartz appeared a fine prism fully 3 inches in length and 3/4 of an inch in its longest diameter. This crystal was of the most perfect transparency and of the purest grass green color, and some of the gems cut from it were almost a match in beauty to the Peruvian emeralds. Another crystal, of unknown length, but more than an inch in diameter, was of a clear blue green in its center, surrounded with a coating of clear white tourmaline a line in depth. This was also surrounded by three other layers of transparent tourmaline, each about a line in depth. The first was pink in hue; the next, limpid white; the last and the exterior was a soft celadon green. There were other crystals of white and green, or white passing to a very light blue. Had the cavity been carefully explored by any one acquainted with mineralogy, the group could have been saved, and a specimen preserved for science surpassing in beauty the valuable mass of pink tourmalines found in Ava, and now to be seen in the British Museum.

The writer, again taking courage at the success of the mica hunters, commenced another exploration on the northern and eastern wall of the pit. Several fine specimens of rose red lepidolite and some other lithia minerals appeared on the side of the excavation, to give hope to the doubting explorer. Eighty tons of rock were removed in this operation before a cavity was struck. Just before the deposit was reached, great masses of lepidolite were found, one of which weighed about five hundred pounds. In fact, it has been noticed often, that the occurrence of lepidolite and smoky quartz are almost sure indications of the existence of a cavity not far distant. The cavity proved to be a large one, of more than a bushel in capacity, and yielded a great number of minute crystals of tourmaline, besides several large specimens, which, unfortunately, were in a state of disintegration. Some months afterwards the exploration was continued, and in the same direction—to the northeast. After removing forty tons of rock, a small cavity (No. 16) the size of the hand was opened, and yielded a broken crystal of dark green, the size of the thumb, and a remarkable slender prism of bluish green, more than 3 inches in length and 1/4 of an inch in diameter. This singular specimen is a perfect facsimile of some of the Siberian beryls, and will readily pass as such. It may be seen in the Vaux collection at Philadelphia. In this last exposure of the ledge, no lepidolite, and very few of the associate minerals that accompany the tourmalines, were obtained; and from the appearance of the wall of the pit, the miners concluded that the eastern limit of the mineral deposit had been reached; therefore the exploration in that direction was stopped. The next summer the western flank was examined; and a few preliminary blasts having yielded positive signs, the miners were directed to blast out an extent of the ledge amounting to about sixty tons. During this removal, several decomposed spots in the albite, enclosing tourmalines, were discovered; and finally a large cavity was reached, which yielded many minute crystals of pure white tourmalines, and fragments of what were once magnificent crystals of white and red, and white and dark blue. A month later in the autumn, the work of blasting was resumed in the same vicinity. Fifty tons of rock were removed; but not a single tourmaline, nor a specimen of the rare minerals associated with them, was brought to light. We then concluded that both flanks of the deposit had been reached, and that the only hope of obtaining further tourmalines lay in blasting out the central portion of the ledge.

Chapter IV

Unwilling to abandon the further search, and urged by mineralogists from all parts of the country, the writer invited the assistance of some of his friends, and formed the Mount Mica Company, and with this assistance the explorations have been continued with a few intervals up to the present day. During 1881 several cavities were opened, and many choice gems and crystals were found. In some of the cavities the crystals of tourmalines were completely disintegrated by the effects of time, and could not be restored. As the tops of the cavities were removed the crystals were disclosed lying in a bed of loose Cookeite in all of their beauty of form and color, but in attempting to remove them some of the most beautiful crumbled away, leaving generally a nodule of clear white or green in the center of the shaft of the crystal, and always of the most perfect transparency. Some of the crystals crumbled away only in the lower parts, leaving the summits or the upper portions of the crystals in perfect condition. Some of these were clear white, tipped with an exquisite shade of green, and so delicately blended that it was quite impossible to tell where the green commenced. The choicest specimens of this year's exploration were two crystals of marvelous beauty, and which are probably the most remarkable known. They were nearly of the

FREDERICK CUTTING HAMLIN

same size and exactly of the same arrangement of color. They are about 3 1/3 inches in length by 7/8 of an inch in diameter, and although broken into three or four parts, they have been restored, and now appear almost as beautiful as when nature created them. The summits of the crystals are not well defined, but are somewhat concoidal in form, and are of the finest hue of grass green. In the middle third of the shaft the green fades into clear white, which in turn changes in the lower third into a fine pink, which gives way at the extreme end of the base to a patch of decided indigo blue. The crystal, viewed from its base along its axis, appears to have a center of blue coated with white, then a layer of pink, and another layer of white coated on the exterior with green. Both crystals are of the finest transparency. (*Vide* frontispiece, plate No. 15.)

In the month of September, 1882, a display of the treasures of Mount Mica was given in the academy hall at Paris, and was attended by many mineralogists and scientists from different parts of the country. On this occasion many of the choicest specimens from the collections of Carter and Hamlin, and others, were exhibited. Besides the cabinet specimens of rare minerals, many of the beautiful gems found at Mount Mica, from the earliest times down to the present, were exhibited. Among them was the remarkable blue tourmaline or indicolite, which was found in cavity No. 26 a few days before by Mr. Carter, and which at the close of the exhibition went into the possession of Mr. Geo. F. Kunz, who has described it with an illustration in color in his splendid work on gems, and therefore we omit giving a drawing of it in this monograph. This was the first public display of the products of Mount Mica, and as an exhibition of the various forms of the lithia group of minerals, it has never been equaled in any country, and may not be again, as since this time many of the rare specimens and gems have been widely separated over the world, and it would be quite impossible to collect them again. For two or three years after this display, but little exploration was done on the ledge, but in the spring of 1886 work was commenced again in earnest, and in May Mr. Carter found a large cavity which promised well, but did not yield many good specimens.

In the month of September, Samuel R. Carter, acting for the Mount Mica Company, removed a part of the ledge in the rear wall of the pit, descending to the depth of twelve feet, and covering a space of about twelve feet square. At this depth he struck one of the largest cavities yet found at Mount Mica, and one of the richest in tourmalines and rare minerals. Until the depth of twelve feet was reached the ledge was barren of interesting minerals, and nothing except a huge black tourmaline four feet in length was brought to light, to indicate the wealth of rare minerals that lay below. About five feet of the overlying rock consisted of a brown mica schist, resting upon the coarse granite vein in which the sheets of mica and the lithia minerals occur. At this depth of twelve feet, masses of quartz, crystals of black tourmaline, and mica appeared to view, and just below them was found a great cavity about four feet square. Along the sides of the cavity, and at the bottom, embedded in the sand of decomposed Cookeite, lepidolite, etc., or lying loose on its floor, were found certainly fifty crystals, or fragments of that number of well defined crystals of tourmalines. They were all of a dark grass green, or blue green, and one of them is the largest crystal known. It is, indeed, a beautiful cabinet specimen, and sufficiently transparent in its middle third, to yield some good gems. It measures 10 inches in length by 2 in diameter, and although broken into four parts, has been joined easily by cement. Both of its terminations are intact, but are not well faceted, and in fact the finest faceted crystals of tourmaline are nearly always small or of medium size. The cavity yielded another large, fine crystal 7 inches in length by 1 inch in diameter, of a rich blue green, and transparent throughout, but greatly marred by numerous flaws. Although broken into eight fragments, it has been restored, with both terminations preserved. Another beautiful crystal of green tourmaline, of the purest water, was 4 inches long by 1/2 an inch in diameter. Most of the crystals were so badly broken that they could not be restored with satisfaction, and were therefore sent to the lapidary, who obtained many fine and valuable gems from them.

After blasting away the walls of this cavity, and only at the distance of four or five feet directly in the rear, another cavity of large size was discovered, and although not containing so many crystals as the former cavity, it proved to be richer in quality and variety of color. (*Vide* plate No. 18.) Two of the crystals were about 2 inches in length by 3/4 of an inch in diameter, but their terminations were so badly fractured and disintegrated that they could not be restored, and were sent to the lapidary, who cut from them two magnificent gems of the finest water and color, of 34 and 28 karats respectively, and also a rich chrysoberyl green gem of 8 karats. The tourmaline of 28 karats, which is absolutely perfect, and one of the finest gems of the mineral known, rivaling the best of the emeralds by artificial light, was purchased by Tiffany & Co., of New York. The other two—the gem of 34 karats, which has a slight imperfection, and the 8 karat of chrysoberyl green—have been placed in the Hamlin necklace of American gems, and will remain there probably for the future. A number of other and smaller gems were cut from the fragments from these two cavities, and were exhibited, with others, and viewed with so much interest at the World's Fair at Paris, in 1888, as to receive special mention from the French Commission.

The most remarkable specimen found in the cavity was a crystal of green, white, pink and blue. (*Vide* plate No. 19.) It was shattered into several parts, but has been restored into a crystal of 5 inches in length by 9-8ths of an inch in diameter. It must have been, in the days of its perfection, about six inches in length, and was then one of the marvels of the mineral kingdom. The summit was too much disintegrated to be restored, but the rest of the upper half of the shaft is of the most perfect transparency and shade of green. The lower half of the shaft changes into white, passing into red and pink, and changing into a tinge of indigo at the base. This was the only specimen in both of these large cavities which showed a tinge of pink. Green and blue green, with a great number of shades, were the predominating tints. Many of the cavities have shown distinctive traits of color, and some of them in a remarkable manner. One cavity may show crystals of all green, or blue, or white, or tints of red, differing from the others.

Chapter V

For three or four years after this exploration little was done at Mount Mica in the way of development, owing to the belief that the deposit did not extend further to the eastward beyond the pit as thus far explored, and to mine deeper beyond the last cavity required considerable money and courage. In 1890 arrangements were made with Loren B. Merrill and L. Kimball Stone, two young men residing in Paris, both ardent, energetic mineralogists, to work the deposit, and they have continued to do so with skill and energy up to the present time. Under their management the ledge is fast disappearing, and the treasures hidden within its depths have been brought to light, year after year, with great interest to the mineralogists who have attended and watched their labors.

In June, 1891, they opened a cavity near the two last found in the bottom of the pit, and it proved to be one of great mineralogical interest. More than thirty distinct crystals were found in this cavity, some of them of great beauty and perfection of color. Nearly all of them were blue, and exhibited by transmitted light a fine tint of sapphire blue. Quite a number of the crystals were entirely blue, but some of them were tipped with a decided shade of green, or rather the blue changed into green near and at the summit of the crystal. One of these crystals was 5½ inches in length by 3/4 of an inch in diameter, with both terminations intact. At the base the color was almost black, becoming clearer along the shaft towards the top; in the middle third, the indigo blue becomes of a beautiful sapphire blue, becoming lighter in tint, and finally changing into a clear and beautiful green at the summit of the crystal. This rare and matchless crystal was broken into four sections, but has been restored to almost its pristine beauty. (*Vide* plate No. 26.) Another crystal of the same arrangement of color was found, a little larger in dimensions, but not so fine in purity of tint or in perfection of form. It is 6 inches in length by 3/4 of an inch in diameter, and is attached to another crystal of indicolite, passing into black, and of unknown length, as some of its sections have been lost in removing the contents of the cavity. No trace of red or white crystals was found in this cavity.

The gems cut from some of the fragments do not exhibit the fine tints of blue that were expected from them, and some of the purest pieces fail to show when cut the beauty of hue that was shown by them before cutting, when examined by a ray of sunlight. The rays of light seem to be absorbed, and do not refract in the usual manner with the expected hues.

In 1891 Loren Merrill and Stone examined the surface of the ledge to the eastward of the pit, with the view of an exploration if a close investigation warranted the expenditure of time and money. About forty feet to the eastward of the pit some Cleavelandite was visible, with black tourmaline and smoky quartz, the combination of which was favorable to the occurrence of tourmaline deposits; and, moreover, this spot was the identical place where Mr.

Hamlin found the first specimen. So the young mineralogists resolved to explore the ledge at this part, and almost the first blast revealed the existence of extensive deposits which had long been unsuspected by the many explorers who had examined the ledge for more than half a century. During this year all the work done at Mount Mica was at this new locality, or rather extension of the deposit, and some very fine specimens were brought to light, the finest of which are drawn among the plates. They were all green, red and white, without any traces of distinct blue. In 1892 work was resumed in May, and many cavities were opened, and some beautiful crystals were found, which are described in the colored plates. During the explorations of this season many barren cavities were opened, and although the surrounding rock was rich in minerals of the lithia group, nature seemed to have forgotten to enrich the cavity with crystals of well-defined tourmalines. This frequency of barren cavities at this region was somewhat remarkable, as in the old pit, which embraced the excavations of seventy years, only two cavities were found that did not contain crystals of tourmalines. In 1893 operations were commenced in July, and some remarkable specimens of tourmalines were discovered, and some very unlike any of those thus far found. A number of them were tipped with red, which is unusual with the crystals of Mount Mica, but common with the tourmalines of Siberia. Some of them have a well-defined zone of indigo blue a line or more in thickness extending across the shaft of the crystals, which are of clear green, both above and below the intrusion of blue. (*Vide* plates.)

Chapter VI

To describe this remarkable deposit in strictly scientific terms will be a difficult task. The ledge in its early days of examination seemed to be foliated, not stratified, and consisted of layers of granite, bending toward the northwest. This inclination of the layers, at first gentle, is now, at the back of the pit, at the depth of sixteen feet, found to be almost perpendicular. These folds of granite lie like the leaves of a book, but not of a definite thickness, and as they bent over to a certain extent, the coarse granite of the upper layers suddenly changed in character. It was granite still, but the arrangement of its particles exhibited a decided change. The masses, flakes and coarse crystals of albite, the large nodules of quartz, the broad plates of mica, and the huge and numerous crystals of black schorl vanished, and instead of them the ledge appeared of firmer texture, but composed of much smaller particles of the same materials. The line of demarcation was quite apparent; yet there was no line of decided and distinct separation. Along this imaginary streak of changed particles occurred the tourmaline deposits. They sometimes appeared in the folds of granite a foot or two above this ill-defined line, but never below it. In all the cavities known to us, and more than eighty in number, we are not aware of one found below this change in the rock. The change in the rock is made clear to the miner by the appearance of minute crystals of garnets, and to explore beneath has been thus far a waste of time and labor. The early explorers found the deposits at the surface, and followed them to the southward, about fifteen feet in distance, where the streak had declined to the depth of six feet below the surface. Since then we have pushed the exploration to a greater distance of sixty feet, where the line of deposit appears at the depth of sixteen feet, with an almost perpendicular inclination, and within a few feet of a bed of brown mica schist.

The small area of about fifteen feet square excavated in the early days by the Hamlin boys, Shepard, and others, was fairly honeycombed with cavities, which seemed to be somewhat connected with each other, but the later explorations were conducted with much uncertainty. There has been no connection with the pockets found at later periods, except in one or two instances, where a small group of cavities occurred together; and explorers have been obliged to grope in the dark, and trust to hazard, in their search for the mineral treasures. Cavities were suddenly found at a considerable distance from the last workings, and when hope of success was nearly abandoned. The appearance of lepidolite was often a sign of success, especially when followed by masses of smoky quartz. And when a broad layer of feldspar or albite was found to be changing into regular and broken flakes, a deposit or cavity might be prophesied with great certainty to occur beneath. Many black tourmalines have been seen in the ledge or in the walls of the cavities, but with the exception of three minute crystals, not one has been found in the cavities among the fine crystals of the colored tourmaline.

The cavities generally were roofed with albite, whilst the sides were composed of limpid or smoky quartz mixed with lepidolite, crystals of tin, spodumene, amblygonite and other rare minerals. These cavities were of irregular shapes, and of sizes extending from the capacity of a pint to that of four or five bushels, and their interior has generally been filled with a substance resembling sand, but which is disintegrated Cookeite and lepidolite. Lying in this sand, and generally at the bottom of the cavity, appeared the beautiful tourmalines, often unattached and unconnected with any matrix. Sometimes they were attached to the walls of the cavity. Occasionally the quartz rock in the walls would contain fine crystals of pellucid or smoky hues, which were often transfixed with slender crystals of tourmalines of various colors. The walls of the cavities, though composed of the strongest materials, were always or often found rent and shattered by some unknown force, and by the same agency the crystals of tourmalines were injured in their structure. Sometimes the shafts of the perfectly crystallized tourmalines were found broken into two or three parts, and in other instances they were fractured into numberless minute fragments, falling into sand when touched by the hand. Nature had evidently made these forms of crystallization in absolute perfection, and the process of decay by some unknown force had happened long afterwards.

From the data thus far obtained, the area of the deposit is quite limited, and apparently does not extend over three hundred feet in length. Its depth is not yet determined, but at the depth of the deepest excavation—sixteen feet—two of the largest and richest cavities were found, and we have reason to believe that a rich field of minerals may be reached before the mica schist interferes. But we do not believe that the tourmaline, with all its perfection of form and color, will be found at any great depths below the surface. Most if not all of the colored gems are found in superficial deposits, like the emerald, the beryl, the opal, the topaz, Chrysoprase, etc., and it seems as though the light of heaven was required to produce the beautiful colors of the gems, as it is for the marvelous and varied hues of the flowers of vegetation. It may be affirmed, perhaps, that the contact of the air or a ray of sunlight is required to build up the forms and perfect the colors of many of the rare minerals.

Among the great variety of minerals found in the limited space of Mount Mica may be enumerated pink, white and gray Lepidolite; pink, red, brown, white, green, blue, yellow and black Tourmalines; white, gray and yellow Cookeite; Albite, Feldspar, Cleavelandite, Apatite, Kaolin; white, amethystine, smoky, yellow and clear, colorless Quartz; white and green Beryl; Biotite, Cassiterite, Columbite, Damourite, Fibrolite, Brookeite, Blend, Childrenite, Garnet, Muscovite, Haloisite, Petalite, Nacrite, Spodumene, Hiddenite, Triphyllite, Uranite, Yttrocerite, Zircon, Tungsten, Autunite Granite, Graphic Granite, Adularia, Montmorillonite, Tantalum, etc.

The color suite of the tourmaline comprises all the tints of the solar spectrum, and is probably the only mineral yet known that exhibits such a vast range of hue, surpassing, probably, even the varieties of transparent corundum or sapphire. It has been stated that none of the vegetable productions exhibit the entire range of the hues of the solar spectrum in the colors of their petals. The tulip, it is said, has the greatest range, but none of its petals are ever black. The Mount Mica tourmalines show a great extent of color, but as yet we have seen none of the true pigeon blood tint, like some we have seen in the tourmaline specimens from Siberia and Brazil, although some of the crimson tints of the Mount Mica stones are of a gorgeous hue.

The arrangement of color in the crystals of tourmaline is also very remarkable, and reminds one of the diverse coloring seen sometimes in the corundum or transparent sapphire, but upon a far more extended scale. In some of the crystals the red changes into blue, and the blue finally passes into green or black; or the red may pass into white, and the white be tipped with green. In others, the color is simply red and green, or white and green, exhibiting many intermediate shades. Generally, these transitions and gradations of color are imperceptible as they pass into each other. But in some specimens the colors are not mingled in the least, and the line of demarcation is well defined and trenchant. So sharply distinct are these crystals in color that they seemed to be composed of several sections veneered together; yet these stones are homogeneous, and cannot be cleaved apart any more than the bands of the onyx. With the tourmalines of this locality, we have noticed that the faceted terminations are nearly always green, while the red portion is generally at the base, which is flat. When the crystals are all red, they are not well terminated, nor well defined in form or prism, at Mount Mica. The Brazilian tourmalines are rarely faceted with perfect planes, no matter what the color may be, while the red tourmalines of Siberia are often beautifully faceted at their terminations.

Masses of gray lepidolite and Cookeite have been observed filled with small crystals of tourmalines, hollow, like thin tubes of glass, with their interior coated, completely or partially, with yellow Cookeite, arranged in filaments, in tufts, or in masses. Some crystals have been found composed of a columnar structure, made up, as it were, with bundles of acicular crystals, which are sometimes drawn out to a delicate fineness; and in several instances they have been seen arranged in groups, and as minute and silken as the thistle's down.

From the evidence collected by or known personally to us, we believe that Mount Mica has yielded more than one hundred crystals which would be considered as fine and remarkable specimens of the mineral. Of the smaller tourmalines, ranging from one inch down to microscopic size, no correct estimate can be made; but they amount to many thousands. We have seen specimens containing more than fifty distinct and transparent crystals, imbedded in masses of lepidolite, Cookeite and albite. Coarse and opaque, or even translucent, crystals of tourmaline, several inches in diameter and nearly a foot in length, have been found in the great masses of albite and quartz; but all the fine and transparent prisms, with but few exceptions, have been taken from the cavities. These exceptions refer to a few crystals found in portions of feldspar, which were soft and partly decomposed, or in deposits of kaolin.

Sometimes the minute crystals may be seen penetrating limpid quartz, like the specimens found in the Ural mountains in Siberia, and cut at Ekaterinsburg into gems and ornamental stones. They then appear like arrows of rutile enclosed in the quartz, but of red and green hues, presenting a beautiful appearance. Well marked specimens of dislocated and curved crystals have frequently been found, and some beautifully radiated tourmalines of a transparent green color—but never red—have been disclosed by rifting masses of mica. And sometimes we observe in the solid masses of quartz or feldspar well defined crystals of tourmaline, articulated like pillars of basalt, and whose sections have been separated at some distance by the intervening rock, as we often see in specimens of beryl. The separation has evidently taken place while the crystal was forming, for the shaft of the prism is often complete and symmetrical, although its sections may be separated at the distance of several inches. This peculiarity is noticed with all of the varieties, and is particularly marked in the black crystals. Many perforated crystals of tourmaline have also been found, occurring in thin, glass-like tubes, sometimes more than an inch in length, but generally less. And the interior of these singular tubes is often free from any substance, or it may be filled with tufts of variously colored Cookeite. Attached to masses of quartz or feldspar, we have often observed singular cavities of small capacity, and whose walls were composed of Cookeite, or a substance resembling Cookeite. The sides of these cavities were sometimes beautifully and clearly striated, as though nature had prepared a mould, and had intended to deposit therein the crystals of tourmalines, but had forgotten to do so, or had removed them by some mysterious law. Some of these cavities were studded on their internal side with minute transparent crystals of quartz, partly covering the clearly defined striæ.

All the crystals of tourmalines found at Mount Mica do not have perfect terminations, and it is very rare that a perfect prism is found; very often we meet with them without any well-fined faces. Those found in masses of kaolin especially are of irregular forms and terminations, indicating that nature, restrained by disturbing causes, has left her work imperfect, both in symmetry and in color. This hiatus or peculiarity in the regularity of the deposition and crystallization of this mineral is far more common with the pale pink tourmalines than with any other variety. Frequently we have observed well defined prisms of tourmaline transfixed by other crystals of the same substance.

The tourmaline is exceedingly interesting to the student, on account of its complex mineralogical characters and curious physical properties, in which respect it far surpasses all the other gems. The inquirer will find much to interest him, if he will turn to the experiments of the German physician, Mr. Æpinus, of the last century, and to those of Mr. Canton, the English electrician, whose researches were published in the proceedings of the Royal Society in 1759. Much of interest also can be found in the curious experiments explained at length in the Philosophical Transactions, Franklin's Letters, and Dr. Priestley's Works. And since these times science has added greatly to these phenomena. Among the curious properties of the mineral in its perfect form is the strange play of color, which is called dichroism or polychroism, when the transparent prism displays two or more colors when viewed in different directions. But few of the gems possess this singular property, even in a slight degree; but in the tourmaline the display of polychroism is seen in its greatest perfection. Some of the prisms of transparent tourmaline, when viewed parallel to their axes, appear of a splendid crimson hue; but when the crystal is slightly turned, the red color vanishes as if by magic, and the clear gem becomes white or smoky in hue, without the least tinge of its former color. Other crystals may exhibit a clear and lovely shade of green when viewed transversely, but the beautiful tints vanish when the same crystal is looked at along its axis, and yellow brown hues appear instead. Some crystals may be dark violet transversely, and greenish blue axially. The range of the diversity of color displayed by this mineral, when viewed in this manner, is very great, but all the crystals or masses do not display this property with equal intensity. Some exhibit it with great distinctness, while other specimens display only a trace of it, and some, none whatever. Turn the fragment however you will, the color remains the same, and unchanged. This absence of dichroism is best observed in the light colored specimens, which also possess the property of double refraction in a feeble degree. The optical characters of this mineral are sometimes wonderful, and some of the prisms, when viewed perpendicular to their sides, appear of a clear and lively color and perfectly transparent; but when they are observed in the direction of their axes, the same limpid stones become perfectly opaque. In some of the specimens, even when the length of the prism is less than its thickness, not a ray of light can be made to glimmer through them, and it is with such that the peculiarity of absorbing one of the rays of polarized light is seen in its greatest perfection.

It is interesting to examine this wonderful mineral deposit at Mount Mica, where the tourmaline occurs in such perfect and wondrous beauty, and to conjecture how nature constructed the marvelous stones in the very heart of the granite rocks; how she silently built up in the darkness of the miniature caverns, or in the very substance of the granite itself, the transparent atoms of their crystal forms; how she touched them with the fiery red, the lively green, the mellow yellow, the somber black, or the tender blue; how, at times, she separated these hues in the same crystals as if by magic touch, or blended them together in exquisite transition and gradation. Here, among this grand display of the rare and the beautiful, Steno might have properly spoken of the play of nature—Steno, who began geology; whom Deluc called the first geologist.

CHAPTER VII

*Explanations of the Plan of the
Excavations of Mount Mica to* 1891.

The space enclosed within the dotted lines represents the area explored from the earliest times up to the period of 1867. It was about fifteen feet long and not over six feet in depth in the deepest part along the rear wall, and it contained nearly all of the cavities found by the Hamlin boys, Profs. Shepard and Webster, and others.

No. 1 was the cavity found by Mr. Bowker before 1866, and which yielded many fine specimens, which have since been scattered and all trace of them lost. No. 2 is the cavity which yielded the crystal of plate No. 7, and was blasted out by A. C. Hamlin in 1868. No. 3 was explored by E. L. Hamlin in 1868. No. 5 by A. C. Hamlin in 1869; also No. 6, in 1869, by E. L. Hamlin. No. 7 was explored by A. C. Hamlin and Prof. Leidy in 1870; also, No. 8, by the same parties, 1871. In this cavity were found the large white crystals of tourmalines. Nos. 9, 10, 11 and 12 were discovered by the men exploring for mica in 1870. No. 13 was opened by A. C. Hamlin in 1879. Nos. 14 and 15 were blasted out in 1878 by Bowker & Perry. No. 16 was opened by A. C. Hamlin and Mr. Vaux in 1879. No. 17 was explored by A. C. Hamlin in 1881.

THE HISTORY OF MOUNT MICA 129

No. 18 by A. C. Hamlin in 1873. Nos. 19 and 20 were examined by Saml. R. Carter in 1881; also, No. 21 in 1882; also, Nos. 22, 23 and 24 by the same explorer in September, 1886. No. 25 was discovered by Merrill & Stone in 1891. No. 26 was opened by S. R. Carter in May, 1886. No. 27 by Bowker & Perry, in 1880.

Besides these cavities enumerated, there were several others found, which proved to be barren, or yielded small results. Since 1891 all of the operations at the ledge have been conducted on the left of this the old pit, and from fifty to one hundred feet from the last workings or from cavity No. 24. The old pit has been abandoned only temporarily, and chiefly because the new pit seemingly offered a richer field with a less amount of blasting.

Explanations of the Plan of the New Pit.

The new pit up to May, 1895, has been excavated about one hundred feet in length, and fifty feet in width, and to the depth in the left corner of about ten feet. Nearly sixty cavities have been opened and numbered, but many of them were vacant, or contained but little of mineralogical interest. Cavity No. 1 contained quite a number of green tourmalines of considerable value. Cavity No. 14 contained the great crystal, the summit of which was found by Miss Hubbard in the soil, and is fully described in plate No. 32. Cavity No. 23 yielded some green crystals of value. Cavity No. 38 also afforded a fine green crystal. Cavity No. 43 contained the splendid crystal which is described in plate No. 43, and which yielded the beautiful gem of fine green hue of the great weight of 69 5-16 karats, and also a superb pink one of 18 karats. In some of the cavities around No. 40 were found the remarkable crystals which exhibited the red tops, with a band of blue in the shaft. In cavity No. 51 was found the valuable crystal which yielded a perfect green gem of 14 karats weight, and also from the same prism a beautiful gem of a fine brown with a crimson tint, of 16 karats weight.

Illustrations.

The plates are intended to be faithful representations, and as near as possible facsimiles, both in color and in form, of the original crystals, either as they were found or as nature made them at the time of their creation. They have been made with great care, under the scrutiny of other mineralogists, and are taken from the original drawings of the author, and produced in color by the Coloritype Company, of New York. No attempt at exaggeration in color has been made, but the degree of hue has been taken by transmitted light, and has been accepted as correct by competent judges. Where reduction has been made on account of the size of the crystal and the limited space of the page, attention has been called to the fact in the description of the plates. This list of illustrations is far from being complete, but it may serve to give the observer and reader an idea of the beauty and the perfection of some of the mineral treasures which nature has deposited at Mount Mica. Many of the choicest crystals herein described have been placed in the mineralogical cabinet of Harvard University by James A. Garland, Esq., of New York, whose thoughtful liberality has also enabled the author to produce the exquisite illustrations of this work.

Explanation of Plates.

Plate No. 1 is a fine crystal of clear grass green color, in two fragments or sections. The upper section, with well defined faces, was found in the soil near the old pit in 1879, while the lower section is the identical fragment which attracted the eye of Elijah L. Hamlin when he discovered the locality, in 1820. For fifty-nine years the summit of the crystal had escaped the observation of the many explorers who had visited the ledge, and dug and blasted about it, during this period of time. The two fragments had become separated during a long period of time by the rains and the winds, and were found several rods apart.—*Hamlin Cabinet.*

Plate No. 2. This is one of the crystals found at Mount Mica in 1825 by Prof. Charles Upham Shepard, of Amherst College, and is described by him in *Silliman's Journal* in July, 1830. The drawing is made from description, and not from actual view.—*Amherst Cabinet.*

Plate No. 3. This represents another of the remarkable find made by Prof. Shepard in 1825, and is drawn from description. Both terminations seem to be absent with these specimens, and probably no attempt was made by Prof. Shepard to restore the crystals to their primitive form with the fragments that had become detached by frost, or other causes, and so we have no means of knowing how large the crystals might have been in their original state, or how diversified their color.—*Amherst Cabinet.*

Plate No. 4 represents a section of a pink crystal found in 1820, and kept in the Hamlin family; some gems of a pink hue were cut from it many years ago. Its green top shows the sharp line of demarcation of color in the same substance. No. 2 represents a remarkable crystal of pink tourmaline found in 1869, by men blasting for mica, and it was found in a mass of kaolin, and nothing of its substance was lost. It appears as nature made it, and several others have been found in kaolin, of singular forms, and quite as difficult to explain as this.—*Hamlin Cabinet.*

Plate No. 5. This is the crystal found by Mrs. Bowker, and given in 1838 to Prof. Charles T. Jackson, when visiting Mount Mica and while making the geological survey of the state. It was found some time before, by Mrs. Bowker, loose in the soil, and at the time of discovery was without a flaw. When last seen by the author it showed three minute flaws, which Jackson thought might have been caused by experimenting with it in hot baths of mercury. It exhibits a light hue of green, with a tinge of blue, terminating at one of the ends in a faint pink. Both of the ends have been polished square by the lapidary. No 2 is a facsimile of the beautiful tourmaline found in a small cavity in 1863 by Prof. Sanborn Tenney. At the meeting of the American Association for the Advancement of Science at Salem, in 1869, this crystal, with the others found in the same cavity, was exhibited to the author, and this sketch was made. It was then of great beauty, and would have yielded a superb gem of a lovely green tint, weighing more than 30 karats, and of great value. Ten years after this examination it was again shown to the author, and during this interval of time a great change had taken place in its appearance, and the specimen had become so disintegrated as to be of little commercial value. The cause of this rapid destruction is unknown. It is now, with other crystals found in the same cavity, in the Yale Cabinet.

Plate No. 6. This figure represents the little crystal found in a small cavity or decayed place behind a lump of lepidolite in the wall of the pit, in August, 1868, and was the cause of subsequent extensive explorations, the deposit at this time having been regarded as exhausted. No. 2 was found in July, 1892, is transparent, of fine color, and although broken into three parts has been easily restored.—*Hamlin Cabinet.*

Plate No. 7. This remarkable crystal of tourmaline was found at Mount Mica in 1868, in cavity No. 7, at the depth of six feet from or below the surface of the ledge. The cavity was scarcely larger than the hand, and contained nothing but this single crystal. It is transparent, and weights 6½ ounces. The green hues are of grass green and the red are of the richest crimson.—*Hamlin Cabinet.*

Plate No. 8. This is the most remarkable crystal of white tourmaline or achroite known. It was found in 1869, in the exploration made with Prof. Joseph Leidy, in cavity No. 8. The cavity was a large one, and yielded several other crystals of smaller size. This crystal is transparent, white at the top, passing into a smoky tinge toward the base, but appears of a crimson hue when viewed along its axis. Its terminations are both tipped with green, and are not well defined or terminated with faces.—*Harvard Cabinet.*

Plate No. 9. This beautiful crystal of transparent tourmaline is restored from fragments rescued from the men blasting for mica in 1869, and represents it in its natural and perfect condition. Found in cavity No. 10.—*Harvard Cabinet.*

Plate No. 10. This singular crystal of tourmaline capped with white was found in the explorations of

1869, in cavity No. 8, and is but slightly restored.—*Harvard Cabinet.*

Plate No. 11. Facsimile of a crystal found in 1870, in cavity No. 7, and sent to the Leidy Cabinet. It is translucent and transparent, with both terminations preserved.

Plate No. 12. No. 1 was found in 1870, and sent to the Leidy Cabinet. It is acicular in structure, and pink and light green in color. No. 2 represents a portion of a large crystal found in the soil in 1879. The central part of the crystal is of a deep blue, passing into black at the base, while at the top it changes into a rich crimson. The outside of the crystal is coated with light green.—*Harvard Cabinet.*

Plate No. 13. This illustration represents one of the finest crystals found at Mount Mica. The first section was found in the soil in 1879, and was cut into gems, one of which, weighing 28 karats, may be seen in the Harvard Cabinet. The other two sections were found after the first had been cut, in cavity No. 27, and the summit is still missing. It was probably about five inches in length, and in its perfect condition was one of the very finest specimens known. It is of perfect transparency, and of the purest grass green, with a slight tinge of blue.—*Harvard Cabinet.*

Plate No. 14. This extraordinary crystal of transparent blue-green tourmaline was found in the rich yield of 1886, in cavity No. 23. It was broken into eight parts, but has been restored, and measures about seven inches in length. Its shaft is too badly flawed to afford any gems, but both terminations have been preserved, making a remarkable specimen. Now in the *Carter Collection.*

Plate No. 15. This wonderful crystal of green, white, pink and blue tourmaline was found in cavity

No. 20, in 1881, and the cavity contained two crystals of the same remarkable coloring, and differed only in size, one being a little larger than the other. The larger one is represented here. It was broken in three parts, but has been easily restored. It is of the purest transparency, and its summit is of the richest grass green, changing toward the base into white, then pink, and finally into indigo blue.—*Harvard-Hamlin.*

Plate No. 16. No. 1 was found in one of the cavities of 1881, No. 19, with several others of similar marking, and its termination is clearly defined. Its color is of an exquisite green, passing into white, and so imperceptibly that it is difficult to tell where it ends. No. 2 is one of several crystals of similar colorings found in the cavities of 1881, and is of fine hue and transparency, but without well-defined terminations.—*Harvard Cabinet.*

Plate No. 17. No. 1 is a crystal found in 1881, in No. 20, and exhibits three distinct colors—red, white and blue; it was broken into three parts, but was easily restored. Neither one of its terminations was well defined. No. 2 is a part of a crystal found in the same cavity, and was probably much longer before the terminations were disintegrated. The top is of a delicate green of the purest water, and fades imperceptibly into white—in fact, no line of demarcation can be distinguished.—*Harvard.*

Plate No. 18. These two sections represent parts of two choice crystals, whose terminations were so disintegrated that restoration was impossible, and no estimate of their original length could be safely determined. These fragments yielded three magnificent gems, which are among the finest known of the species. One of them was a perfect stone of the finest green, of 27½ karats weight, and is now in the Tiffany collection; the other was of the same exquisite shade of green, but had a minute flaw. It weighed 34½ karats, and is now in the American necklace of native gems in the Hamlin Cabinet. The top of the longer section furnished also a beautiful stone of chrysoberyl green—cut parallel to its axis—of about 7 karats, and was exhibited at the Paris Exhibition, and is now in the Hamlin necklace. Found in cavity No. 24.

Plate No. 19. This is one of the most wonderful crystals of tourmaline that Mount Mica or any other country has yet produced. It was found in the cavities of 1886, No. 24, and its six broken parts were taken out with care. Its top was too much disintegrated to be replaced, but the other parts have been restored as it appears in the plate. The top of the crystal is of the purest green and of the finest water, with but few flaws; the center is white, passing into pink toward the base, which terminates in a decided indigo blue.—*Harvard Cabinet.*

Plate No. 20. These four green crystals of irregular forms are of perfect transparency and purest color, and were found in 1879, in cavity No. 13.—*Hamlin and Harvard Cabinets.*

Plate No. 21. This beautiful crystal of blue tourmaline or indicolite was found by Fred. C. Hamlin in a small cavity, in 1881. It is of perfect transparency and of a fine blue color, tinged slightly with green. Although broken into four parts, it has been united with both terminations preserved.—*Hamlin Cabinet.*

Plate No. 22. This remarkable crystal of blue tourmaline was found in the explorations of 1891, in cavity No. 26. The illustration shows it as it appears by transmitted light. The upper third of its shaft exhibits a rich, transparent sapphire blue, changing into opaque black at its base. Attached to another crystal of blue-black.—*Carter Collection.*

Plate No. 23. This crystal, of the finest transparent indigo blue, passing into green at the summit, was found in cavity No. 26, in 1891, and as it had but one termination it was cut into gems.

Plate No. 24. Transparent indicolite, found in cavity No. 26, broken into four parts, restored with both terminations preserved.

Plate No. 25. Restored crystal of indicolite, with both terminations; color a fine blue, passing into green at the summit. Cavity 26, 1891.—*Harvard Cabinet.*

Plate No. 26. This, a representation of the finest crystal of indicolite thus far known, is transparent throughout its entire shaft, and although broken into five parts, has been successfully restored with both of its terminations complete; color a beautiful sapphire blue, changing into a delicate green at its top.—*Hamlin Cabinet.*

Plate No. 27. Restored crystal of blue and green tourmaline, of fine tints, with both terminations intact, found in 1891.—*Harvard Cabinet.*

Plate No. 28. These two crystals of indicolite were found in June, 1891.

Plate No. 29. These two rare crystals of blue and green tourmaline, with terminations complete, and of the finest transparency, were found in 1891, in cavity No. 26.

Plate No. 30. This illustration represents the largest transparent crystal of green tourmaline known, and is one-half the natural size. It was found in September, 1886, in cavity No. 23, by Samuel R. Carter, Esq., while exploring for the Mount Mica Company.

It is 10 inches in length and 2¼ inches in diameter, and weighs 41 ounces. Both terminations have been preserved, but they are not well defined.—*Harvard*.

Plate No. 31. The illustration is three-fourths the size of the natural crystal, which was found in October, 1891, in the new pit, by Merrill and Stone, while mining for the Mount Mica Company. It was of fine grass green color, and transparent throughout. It had been badly shattered by the elements, but all of the fragments were adjusted before the sketch was made. The summit is well defined, and has been preserved, but the rest of the shaft has been cut into gems.

Plate No. 32. This is a facsimile of the superb crystal found by Merrill and Stone, in October, 1891, in the new pit, while operating for the company. The summit of the crystal was first discovered by Miss Lizzie Hubbard, of Paris, while carelessly digging in the soil, not far from the miners then blasting in the solid ledge. The attention of the miners was soon called to the spot, and the remainder of the crystal quickly unearthed. The lower half of the crystal was badly fractured, but all the fragments were found and the restoration of the crystal made complete for the drawing. A large number of gems of fine water and of the various colors of green, white, red and pink were cut from it.

Plate No. 33. This represents three-fourths of the natural size of a large green crystal, found by Merrill and Stone, in September, 1892, in the new pit. It was badly broken, but all the fragments were found and the crystal restored and sketched. The summit was partly defined, but the base was without regular form. The crystal was transparent, of a clear grass green hue, with the exception of the extreme base, which was of a decided salmon tint. Cut into gems.

Plate No. 34. Picture of a remarkable curved and dislocated tourmaline, found in 1891. Green in color, transparent, and translucent in places.—*Carter Collection.*

Plate No. 35. This beautiful crystal of various tints was found in July, 1892, and has been preserved. Both terminations complete.

Plate No. 36. This beautiful complete crystal of green, with a tinge of pink at the base, was found in July, 1892. Its upper portion is almost without flaws, and is of superb color and of the finest water.

Plate No. 37. This is a choice crystal of green tourmaline, with a touch of pink at the base, and was found in September, 1891. It was broken into six sections, but was easily restored and preserved. It will yield several fine gems of pure green.

Plate No. 38. Crystal of light bluish green, found in June, 1893, by Merrill and Stone. Terminations not well defined.

Plate No. 39. Crystal of green, tipped with red, found in June, 1893. Transparent and translucent. Broken into three pieces.

Plate No. 40. Drawing of a restored crystal of green and red and white tourmaline, found in June, 1893, in the new pit. It was badly fractured into nine sections, and was restored with some difficulty, so as to make a correct drawing. The summit is well terminated, but the base is not. The extreme top is red, and below the cap appears a band of clear green, followed by red, again passing into white, and then into deep green at the base. The exterior of the crystal is coated with green.

Plate No. 41. The drawings represent two remarkable crystals of tourmalines found in November, 1893. They are tipped with red, then follows a section of clear green, and then a band of indigo blue a line or more in depth, but extending across the entire shaft of the crystal; below this blue, the shaft becomes green again, and at the base of one of them the color changes into a deep green, with a sharp line of demarcation.—*Hamlin.*

Plate No. 42. The illustration represents two remarkable crystals found in November, 1893. The terminations of both of them are intact and the summits well defined and red in color, while the shafts are of faint green, pink or blue.—*Hamlin.*

Plate No. 43. The drawing represents a very remarkable crystal, found in November, 1893. Like the preceding crystal, it was tipped with red; the shaft below was of the purest green, and has furnished the largest gem of its species known. It is of a lively green color, of the finest water, is without a blemish, weighs 69¼ karats, and is now in the Tiffany collection. The shaft below the part yielding the green gem became white or of a faint green, and also afforded some stones of those tints. The shaft then became red, of various shades from a faint pink to a deep carmine red, and also yielded a number of choice gems of those hues, one of which, a pink one, weighed 18 karats. The upper half of the crystal was sound in structure excepting a few cracks which marred the exterior, and here and there extended into the central portion, but the lower half was too badly broken to warrant an attempt to preserve the specimen, and therefore it was placed in the hands of the lapidary.

All of the crystals and fragments found since May, 1891, up to May 1, 1895, have been taken from the new pit or excavations, fifty feet and more to the eastward from the old and historic pit.

PLATE I

Plate II

PLATE III

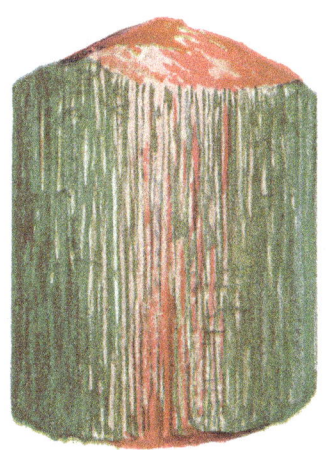

PLATE IV

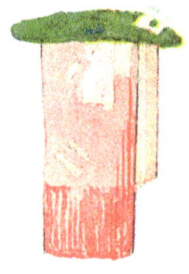

2

1

PLATE V

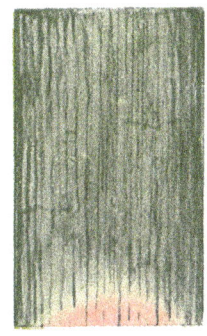

1

2

Plate VI

1.

2

PLATE VII

PLATE VIII

Plate IX

Plate X

PLATE XI

PLATE XII

Plate XIII

Plate XIV

PLATE XV

Plate XVI

2

1

PLATE XVII

PLATE XVIII

PLATE XIX

Plate XX

PLATE XXI

Plate XXII

Plate XXIII

PLATE XXIV

Plate XXV

PLATE XXVI

Plate XXVII

PLATE XXVIII

Plate XXIX

2

1

Plate XXX

Plate XXXI

PLATE XXXI A

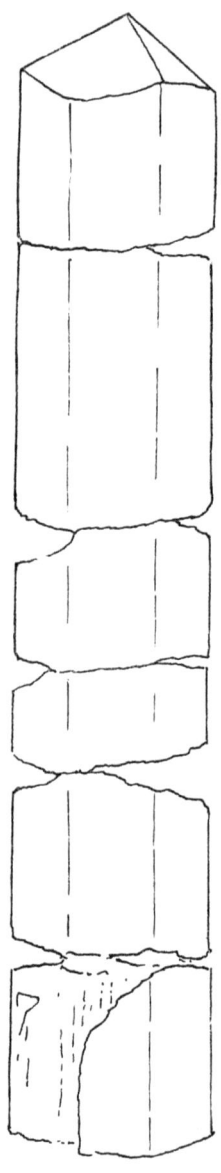

Plate XXXII

PLATE XXXII A

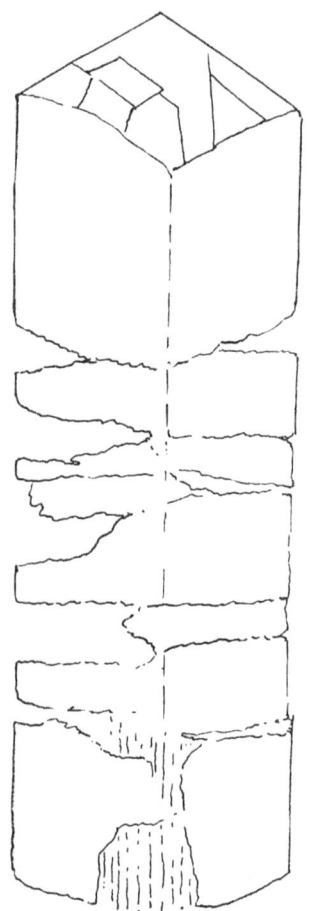

PLATE XXXIII

PLATE XXXIII A

Plate XXXIV

PLATE XXXV

Plate XXXVI

Plate XXXVII

Plate XXXVII A

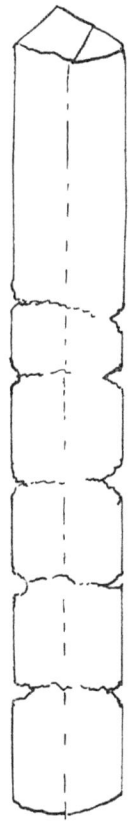

Plate XXXVIII

PLATE XXXIX

PLATE XL

PLATE XL A

PLATE XLI

PLATE XLII

PLATE XLIII

PLATE XLIII A

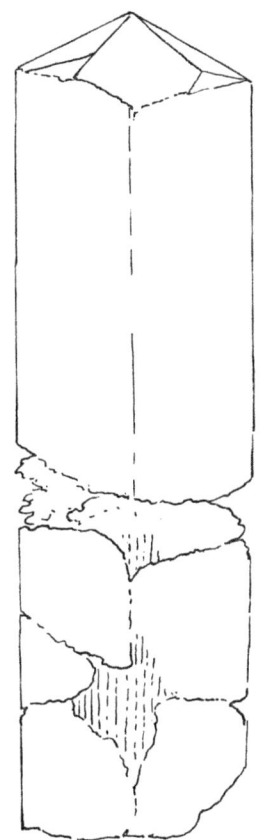

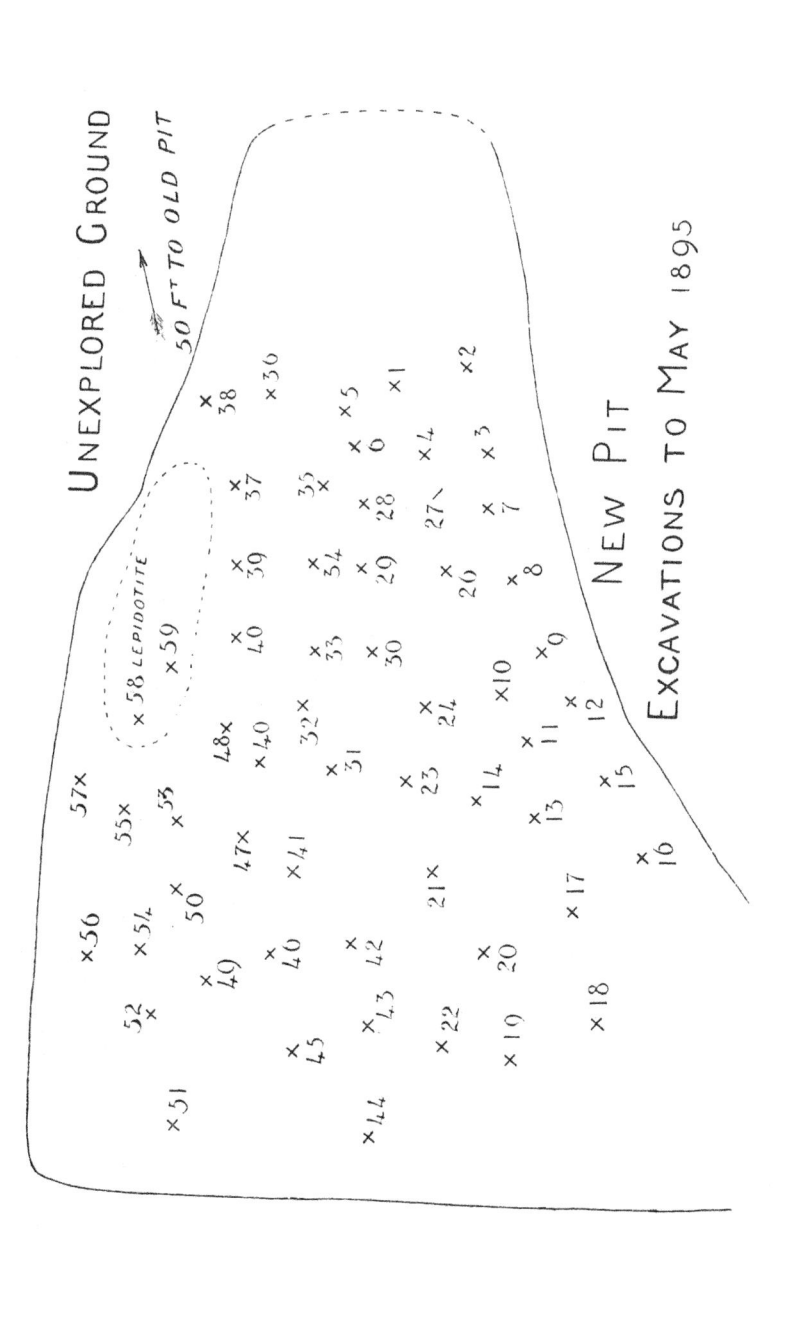

Coachwhip Publications
CoachwhipBooks.com

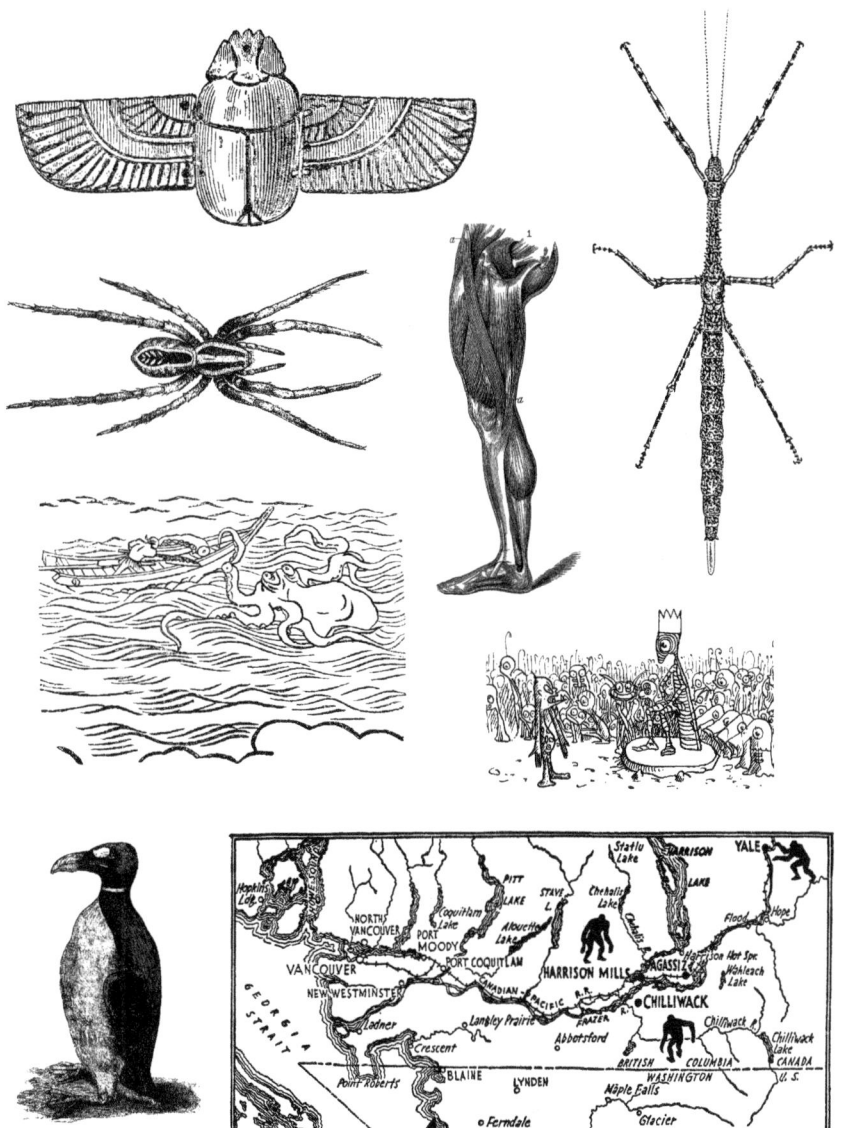

 www.ingramcontent.com/pod-product-compliance
Ingram Content Group UK Ltd.
Pitfield, Milton Keynes, MK11 3LW, UK
UKHW061222180426
11947UKWH00026B/1966